## 权威·前沿·原创

皮书系列为
"十二五""十三五"国家重点图书出版规划项目

三江源绿皮书
GREEN BOOK OF
SANJIANGYUAN REGION

# 三江源生态保护研究报告
（2017）

A RESEARCH REPORT ON SANJIANGYUAN ECOLOGICAL
PROTECTION (2017)

## 水文水资源卷

三江源生态保护基金会
主　编／魏加华

社会科学文献出版社
SOCIAL SCIENCES ACADEMIC PRESS (CHINA)

### 图书在版编目(CIP)数据

三江源生态保护研究报告.2017.水文水资源卷/魏加华主编.--北京：社会科学文献出版社，2018.3
（三江源绿皮书）
ISBN 978-7-5201-1350-2

Ⅰ.①三… Ⅱ.①魏… Ⅲ.①区域生态环境-环境保护-研究报告-青海-2017 Ⅳ.①X321.244

中国版本图书馆CIP数据核字（2017）第220915号

### 三江源绿皮书
### 三江源生态保护研究报告（2017）
——水文水资源卷

主　　编 / 魏加华

出 版 人 / 谢寿光
项目统筹 / 任文武
责任编辑 / 高振华

出　　版 / 社会科学文献出版社·区域发展出版中心（010）59367143
　　　　　　地址：北京市北三环中路甲29号院华龙大厦　邮编：100029
　　　　　　网址：www.ssap.com.cn

发　　行 / 市场营销中心（010）59367081　59367018
印　　装 / 三河市东方印刷有限公司

规　　格 / 开　本：787mm×1092mm　1/16
　　　　　　印　张：12　字　数：175千字
版　　次 / 2018年3月第1版　2018年3月第1次印刷
书　　号 / ISBN 978-7-5201-1350-2
定　　价 / 98.00元

皮书序列号 / PSN G-2018-700-1/1

本书如有印装质量问题，请与读者服务中心（010-59367028）联系

▲ 版权所有　翻印必究

# 《三江源生态保护研究报告（2017）》编辑委员会

**主 任** 王光谦

**顾 问** 王光谦 姚洪仲

**主 编** 魏加华

**副主编** 李铁键

**委 员**（按姓氏拼音排序）

曹 军 陈 刚 傅 汪 黄睿军 黄跃飞

李凤霞 刘 弢 刘锡宁 马志强 苏海红

王志民 王忠静 张贺全 郑桂云

**执笔人** 傅 汪

# 主要著者简介

**魏加华** 男，博士，1971年生，教育部长江学者特聘教授。2004年至今在清华大学水利水电工程系工作，2007年12月至2009年1月为美国康奈尔大学访问学者。2014年9月由清华大学对口支援青海大学，现任青海大学水利电力学院常务副院长，省部共建三江源生态与高原农牧业国家重点实验室水文学及水资源方向首席科学家。

主要研究方向为水文学及水资源、水信息学、流域水量调度、地表水—地下水联合调蓄、空中水资源开发利用等。主持和参与国家自然科学基金、国家科技支撑计划、重点研发计划、公益性行业科研专项、重大工程咨询等课题40余项，发表学术论文150余篇，合编专著5部。先后获得中组部、人事部、中国科协第十届"中国青年科技奖"，国家科技进步二等奖1项，省部级科技进步一等奖5项、二等奖3项。

**李铁键** 男，博士，1981年生，清华大学水利系河流研究所副研究员，教育部长江学者奖励计划青年学者，青海大学水利电力学院昆仑学者讲座教授。分别于2003年和2008年在清华大学水利系获学士学位和博士学位。2010年起在清华大学任教，2014年起参加青海大学对口支援工作。

主要从事河流动力学和水信息学领域的研究工作，成果涉及流域泥沙动力学模型、流域模拟技术、河网几何学、空中水汽通道和空中水资源分析、河流大数据等方面。建立了黄河数字流域模型和数字流域平台，提取了全球高分辨率数字河网 Hydro30，开展了河流大数据方向的探索性研究。发表学术论文 50 余篇，其中 SCI 收录 20 余篇，SCI 引用 200 余次。2015 年获教育部科技进步二等奖（第 2 完成人）；专著《流域泥沙动力学模型》于 2010 年获第二届中国出版政府奖图书奖。IEEE 等学术组织会员，*Water Resources Research, Journal of Hydrology, Hydrological Processes* 等期刊审稿人。

# 摘 要

三江源是长江、黄河和流经六国的澜沧江的发源地,位于我国青海省南部、青藏高原东北部,北纬31°39′~36°12′、东经89°45′~102°23′之间,平均海拔为3500~4800米。按三江源国家生态保护综合试验区划定的行政区域计,三江源包括玉树、果洛、海南、黄南4个藏族自治州全部行政区域的21个县和海西蒙古族藏族自治州格尔木市的唐古拉山镇,总面积为39.5万平方公里。按流域边界计,三江源内黄河流域面积12.20万平方公里(唐乃亥水文站以上),占39.0%;长江流域面积13.77万平方公里(直门达水文站以上),占44.0%;澜沧江流域面积5.29万平方公里(昌都水文站以上),占16.9%。

三江源是河流的源区、水资源的源区,三江源生态保护离不开源区特有水文水资源条件的支撑。因此,在绿皮书的出版计划中,水文水资源被列为首卷。本卷分总报告和分报告两个部分。

总报告首先介绍了三江源的地理位置,厘清了三江流域源区、三江源国家自然保护区、三江源国家生态保护试验区、三江源国家公园等三江源不同地理区域概念及其由来,将范围最广的三江源国家生态保护综合试验区选为本书的研究对象,并以三江流域源区作为参照;随后简要介绍了三江源区的地形地貌特点、气候特征、地表覆盖特征和生物群落情况;最后从人口和产业方面介绍了三江源区的社会经济发展情况。

分报告《水文水资源状况》从降水特征、河流水系、湿地资源、冰川雪山、地下水资源等方面展开介绍。由于三江源区地面雨量站稀少，本文利用卫星遥感降水数据在空间覆盖上的优势，综合采用地面站点数据和卫星遥感数据进行了降水的时间、空间分布特征分析，并对卫星遥感数据进行了对比验证。河流水系分析主要采用清华Hydro30数字河网数据，分别对黄河源区、长江源区、澜沧江源区和内流区的河流、湖泊组成和主要支流概况进行说明，并介绍了科考测量的部分河流断面情况和主要水文站的历史实测径流量，并对湿地资源、冰川雪山、地下水资源进行了简要介绍。

《三江源生态环境保护工作历程及成效》介绍了三江源自然保护区、生态保护综合试验区和国家公园的设立过程，从林草植被、水源涵养、生物多样性等方面介绍了三江源保护措施实施以来的显著成效，并对三江源生态保护综合试验区二期规划和社会各界的三江源保护行动进行了介绍。

三江源区生态保护与建设取得了明显成效，生态系统结构得到局部改善，草地退化趋势受到初步遏制，草畜矛盾趋缓，湿地生态功能逐步提高，湖泊水域面积明显扩大，严重退化区的植被覆盖率明显提升。但也要看到三江源保护仍然面临诸多挑战与问题。

《三江源区生态环境保护工作展望》简述了三江源保护面临的挑战与问题，并从实施好三江源生态保护和建设二期工程等规划、进一步加大科技投入和科学研究力度、建立全社会参与三江源区保护的机制等方面为三江源保护工作提出了建议。

# 目 录

前　言 ………………………………………………………………… 001

## Ⅰ 总报告

**G.1** 三江源区自然特点和社会经济特征 ……………………………… 001
　　一　地理位置 ……………………………………………………… 001
　　二　地形地貌 ……………………………………………………… 005
　　三　气候特征 ……………………………………………………… 006
　　四　地表覆盖特征 ………………………………………………… 007
　　五　生物群落 ……………………………………………………… 008
　　六　社会经济发展 ………………………………………………… 009

## Ⅱ 分报告

**G.2** 水文水资源状况 ………………………………………………… 011
　　一　降水特征 ……………………………………………………… 011

　　二　河流水系 …………………………………………… 018
　　三　湿地资源 …………………………………………… 039
　　四　冰川雪山 …………………………………………… 043
　　五　地下水资源 ………………………………………… 045

G.3　三江源生态环境保护工作历程及成效 …………………… 047
　　一　三江源综合试验区和国家公园的设立 …………… 047
　　二　自然保护区总体规划实施后成效明显 …………… 050
　　三　三江源生态保护综合试验区二期规划 …………… 055
　　四　社会各界在行动 …………………………………… 060

G.4　三江源区生态环境保护工作展望 ………………………… 065
　　一　三江源保护面临的挑战与问题 …………………… 065
　　二　实施好三江源生态保护和建设二期工程规划 …… 067
　　三　进一步加大科技投入和科学研究力度 …………… 069
　　四　建立全社会参与三江源区保护的机制 …………… 071

参考文献 ………………………………………………………… 073

鸣　谢 …………………………………………………………… 074

皮书数据库阅读**使用指南**

# 前　言

"问渠哪得清如许，为有源头活水来。"位于青藏高原的三江源区是我国长江、黄河和流经六国的澜沧江的发源地，该区域每年向三条江河的中下游供水近 400 亿立方米，其中长江总水量的 1.3%、黄河总水量的 38%、澜沧江出省水量的 15% 均源于此，是三条江河的重要水源补给区，素有"江河之源""中华水塔""亚洲水塔"之称。该区域同时也是世界高海拔地区生物多样性最集中、面积最大的地区，是亚洲、北半球乃至全球气候变化的敏感区和启动区，是中国乃至世界生态安全屏障极为重要的组成部分。三江源不仅是中国与东南亚的江河之源、数亿人口的生命之源，更是整个流域经济社会发展的动力之源，因此正确认识和加强三江源区的保护，具有十分重大的意义。

三江源区历史上曾是水草丰美、湖泊星罗棋布、野生动物种群繁多的高原草原草甸区，被称为生态"处女地"。人类文明兴衰于水，正是因为三江源区为我们的母亲河长江、黄河提供了稳定的水源供给，才使得华夏文明得以延续传承 5000 年，至今依然巍然屹立于世界的东方，继续走向强大和复兴。19 世纪 90 年代末至 20 世纪初，全球气候变暖，冰川、雪山逐年萎缩，直接影响高原湖泊和湿地的水源补给，众多的湖泊、湿地面积缩小，甚至干涸，沼泽地消失，导致生态环境愈加脆弱。随着人口增加和人类生产活动影响加剧，该地区生态环境恶化大大加速。三江源区植被与湿地生态系统破坏，水源涵养能力减退，已对我国的供水安全、生态安全和可持续发展构成了巨大

威胁，引起社会各界的高度关注。

党中央、国务院高度重视三江源区的生态保护和建设，中央领导多次做出重要指示、批示。2016年，习近平总书记视察青海省时强调，青海生态地位重要而特殊，必须担负起保护三江源、保护"中华水塔"的重大责任。要坚持保护优先，坚持自然恢复和人工恢复相结合，从实际出发，全面落实主体功能区规划要求，使保障国家生态安全的主体功能得到全面加强。要统筹推进生态工程、节能减排、环境整治、美丽城乡建设，加强自然保护区建设，搞好三江源国家公园体制试点，加强环青海湖地区生态保护，加强沙漠化防治、高寒草原建设，加强退牧还草、退耕还林还草、三北防护林建设，加强节能减排和环境综合治理，确保"一江清水向东流"。

1999年，中国探险协会组织专家对澜沧江进行综合考察，提出了"开发大西北，保护三江源"的建议。这一建议得到国家林业局、青海省政府、中国科协、中国科学院和有关部门的重视与支持。2000年3月，国家林业局、中国科学院和青海省人民政府联合召开了"青海三江源自然保护区可行性研讨会"，青海省人民政府于2000年5月批准建立了三江源省级自然保护区。2005年，国务院批准实施《青海省三江源自然保护区生态保护和建设总体规划》，被称为"新世纪中国生态1号工程"的三江源生态保护和建设工程正式启动。2011年11月，国务院决定在三江源区建立"国家生态保护综合试验区"，并批准通过了《青海三江源国家生态保护综合试验区总体方案》，这是我国第一个以生态保护为主题的综合试验区，我们会努力将该地区建设成为生态文明、经济发展、生活富裕、社会和谐的先行区和示范区。2014年，三江源国家生态保护综合试验区建设暨三江源生态保护和建设二期规划工程正式启动。

# 前言

三江源保护是一项意义重大、使命艰巨、任务繁重的系统工程，同时也是一项惠及三江流域乃至全国人民的宏大工程，它不仅关系到三江源区人民的利益，更关系到我国经济社会的可持续发展、生态文明建设和中华民族伟大复兴"中国梦"宏伟目标的实现，与每一位中华儿女都息息相关，加强三江源保护是我们每个人的重要历史责任和应尽义务。为进一步调动社会各界关心、关注、支持三江源保护的积极性，为全社会搭建支持三江源生态保护的平台，2012年青海省委省政府成立了三江源生态保护基金会。基金会成立以来，多次开展促进三江源生态保护的募捐活动，接受海内外热心环保公益事业的有关组织和个人的捐助，基金主要用于资助三江源资源与生态保护项目和科学研究、科技开发，开展和促进三江源生态保护相关的宣传教育、学术交流以及国际交流与合作等活动。

在《青海省三江源自然保护区生态保护和建设总体规划》的引领下，通过各地各部门通力协作，以及社会各界的共同努力，三江源区生态保护与建设取得明显成效，生态系统结构得到局部改善，草地退化趋势得到初步遏制，草畜矛盾趋缓，湿地生态功能逐步提高，湖泊水域面积明显扩大，流域供水能力明显增强，严重退化区的植被覆盖率明显提升。但同时也要看到，三江源生态环境状况依然堪忧，草地退化、雪线上升、冰川萎缩趋势远未得到彻底遏制，生态保护依然任重道远，还需采取更为科学、全面、细致的综合治理措施，还需动员更广泛的力量参与到"中华水塔"的保护和抢救中来。

追根溯源，只有正确了解认识三江源，才能更好地保护发展她。对三江源的了解和认识，尤其对源头的探寻由古延续至今，从《山海经》描述的"昆仑之丘，河水出焉"、《尚书·禹贡》中的"岷山导江"，到徐霞客在《江源考》一书中所述金沙江是长江的源流，再到

1978年夏，长江源科考队确定沱沱河、南源当曲、西源楚玛尔河的"三源说"，直到2008年科考队还在利用卫星定位系统、遥感等技术对三江源头进行研究。这种对源头的持续探寻，展现的正是中华民族对真理追寻孜孜不倦、实事求是的精神。

三江源绿皮书正是秉承这种精神，利用最新的科技手段和对大量数据资料的校核，对三江源的自然环境、水文资料等进行真实、准确的描述，力求让关心和支持三江源保护的广大人士对三江源区的现状有所了解，也希望能够对政府决策、科学研究等起到参考作用。在王光谦院士的倡议和带领下，三江源生态保护基金会组织编写了该绿皮书。根据基金会的计划安排，绿皮书计划分五册，分别对三江源区的水资源、生态环境保护和发扬三江源区的传统文化、土地利用、三江源生态保护与社会经济发展进行编写。

绿皮书尚有不足处，望各方多提宝贵意见，以期完善。

三江源绿皮书编委会

二〇一七年七月

# 总 报 告
General Report

## G.1 三江源区自然特点和社会经济特征

### 一 地理位置

三江源区位于我国西部、青藏高原腹地、青海省南部。但根据划分依据的不同，三江源区的具体内涵及相应的地理范围有所不同，主要包括"三江流域源区""三江源国家自然保护区""三江源国家生态保护综合试验区""三江源国家公园"等。由于目前社会各界对"三江源区"的具体含义并未达成共识，所以在不同的学术论文和新闻报道中，所涉及的"三江源区"各不相同却又不予以说明，容易对读者造成误导。因此，本文首先对三江源区的几种主要含义进行整理和说明，以示区别。

#### （一）三江流域源区

按照自然流域划分，三江源区是我国黄河、长江、澜沧江三条大江的源头流域，流域面积共31.26万平方公里。其中，黄河流域面积12.20万平方公里（唐乃亥水文站以上，流域面积数据来源于《中国水利统计年鉴》），约占39.0%；长江流域面积13.77万平方公里（直门达水文站以上），约占44.0%；澜沧江流域面积5.29万平方公里（昌都水文站以上），约占16.9%。三大流域源区范围如图1.1所示。

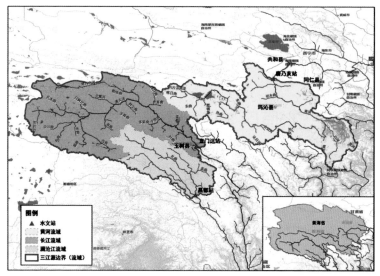

图1.1 按自然流域划分的三江源区范围示意（31.26万平方公里）

## （二）三江源国家自然保护区

为了保护好三江源区的生态环境，2000年5月，青海省政府批准建立三江源自然保护区；2003年1月，国务院正式批准三江源自然保护区晋升为国家级自然保护区；2005年，国务院批准实施《青海省三江源自然保护区生态保护和建设总体规划》，从三江源区域范围内重点选取了15.23万平方公里作为三江源国家级自然保护区，涉及果洛藏族自治州、玉树藏族自治州、黄南藏族自治州的16县1乡，共69个乡镇。按照功能区划分为核心区、缓冲区和实验区，分别占保护区面积的 20.5%、25.8% 和 53.7%。

### （三）三江源国家生态保护综合试验区

2011 年，国务院第 181 次常务会议批准实施的《青海三江源国家生态保护综合试验区总体方案》，划定了三江源国家生态保护综合试验区，总面积达 39.50 万平方公里。行政区划包括青海省玉树藏族自治州、果洛藏族自治州、海南藏族自治州、黄南藏族自治州全部行政区域的 21 个县和格尔木市的唐古拉山镇，共 158 个乡镇。总体方案将三江源试验区划分为重点保护区、一般保护区和承接转移发展区，分别占试验区总面积的 50.1%、47.9% 和 2.0%。三江源国家生态保护综合试验区范围如图 1.2 所示，图中也绘出了按自然流域划分的三江源流域边界，以供对比。

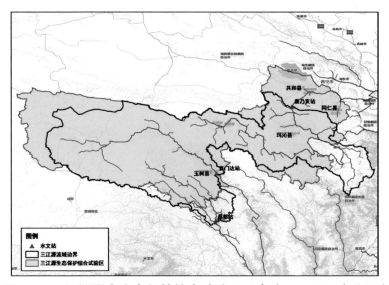

图 1.2　三江源国家生态保护综合试验区示意（39.5 万平方公里）

## （四）三江源国家公园

三江源国家公园包括长江源（可可西里）、黄河源、澜沧江源三个园区，即"一园三区"，涉及果洛藏族自治州玛多县以及玉树藏族自治州杂多、治多、曲麻莱3县和青海可可西里国家级自然保护区管理局管辖区域。园区总面积为12.31万平方公里，占三江源国家生态保护综合试验区总面积的31.2%。其中，冰川雪山833.4平方公里、河湖湿地29842.8平方公里、草地86832.2平方公里、林地495.2平方公里。三江源国家公园的范围如图1.3所示，图中也绘出了三江源国家生态保护综合试验区的范围，以供对比。

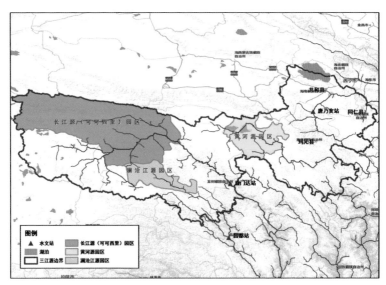

图1.3 三江源国家公园范围示意（12.31万平方公里）

长江源（可可西里）园区：位于玉树藏族自治州治多、曲麻莱两县，包括青海可可西里国家级自然保护区、青海三江源国家级自然保护区索加-曲麻河保护分区，面积9.03万平方公里。涉及治多县索加乡、扎河乡和曲麻莱县曲麻河乡、叶格乡，15个行政村。

黄河源园区：位于果洛藏族自治州玛多县境内，包括青海三江源国家级自然保护区的扎陵湖－鄂陵湖和星星海两个保护分区，面积1.91万平方公里。涉及玛多县黄河乡、扎陵湖乡、玛查理镇，19个行政村。

澜沧江源园区：位于玉树藏族自治州杂多县，包括青海三江源国家级自然保护区果宗木查、昂赛两个保护分区，面积1.37万平方公里。涉及杂多县莫云、查旦、扎青、阿多、昂赛5个乡，19个行政村。

本书将"三江源国家生态保护综合试验区"作为研究区，除特别注明外，后续篇幅提到的"三江源区"特指"三江源国家生态保护综合试验区"，不再重复说明。

## 二 地形地貌

三江源区地处地球上最年轻的高原——青藏高原腹地。这里在两亿年前是海洋，后隆起成为大陆，到几百万年前大幅度隆起，形成青藏高原，至今还在继续隆起。三江源区以山原和峡谷地貌为主，山系绵延，地势高耸，地形复杂，海拔为3335～6564米，平均海拔约4000米。海拔4000～5800米的高山是地貌的基本骨架，主要山脉为东昆仑山及其支脉阿尼玛卿山、巴颜喀拉山和唐古拉山。

三江源区内部各源区因所处地理位置、海拔高程差异，地貌有所不同，长江、澜沧江源区群山耸立，雄伟壮观，以冰川地貌、高山地貌、高平原丘陵地貌为主，间有谷地、盆地及沼泽星罗棋布。黄河源区海拔相对较低，主要是高原低山丘陵地貌、湖盆地貌及河谷地貌。

源区中西部和北部呈山原状，地形起伏不大，多宽阔而平坦的滩地，为高寒草甸区，大面积沼泽湿地；东南部为高山峡谷地带，河流切割强烈，地形破碎，地势陡峭，坡度多在30°以上，相对高差在1000米以上，呈原始森林片状分布；东北部黄河干流自兴海县唐乃亥以下，海拔由3306米（兴海县）逐渐降低到1960米（尖扎县），地势趋于平缓，峡谷、盆地、湿地、阶地相间。

## 三　气候特征

三江源区的气候属青藏高原山地气候，为典型的高原大陆性气候。表现为冷暖两季交替、干湿分明、水热同期；年温差小、日温差大；日照时间长、辐射强烈；植物生长期短，无霜期短或无绝对无霜期。

年平均气温为-5.6℃~3.8℃，其中最冷月（1月）为-13.8℃~-6.6℃，极端最低温度为-48℃，最热月（7月）平均气温为6.4℃~13.2℃，极端最高温度为28℃。基于中国气象局最新公布的中国地面气象数据集，该区域年平均气温在近几十年来呈明显的上升趋势。1981~2000年各气象站点年平均气温的泰森平均值为-1.52℃，2010~2013年各站点年平均气温的泰森平均值为-0.45℃，1981~2013年年平均气温变化率约为0.32℃/10年。2000年之前的年平均气温大于0℃的平均海拔为3646米，2000年后的年平均气温大于0℃的站点新增了达日站，其海拔为3968米，反映本区气候变暖过程中，年平均气温0℃的海拔也呈上升趋势。

该区域多年平均降水量为406.6毫米，其中5~9月降水量约占全年降水量的85.3%，夜雨量比例达55%~66%，年蒸发量在

730～1700毫米。三江源区年日照时数2300～2900小时，年太阳辐射量5658～6469兆焦耳/平方米，全年大于8级大风日数3.9～110天，空气含氧量仅相当于海平面的60%～70%。冷季为长达7个月的青藏冷高压控制，热量低、降水少、风沙大；暖季受西南季风影响产生热气压，水汽丰富、降水较多、夜雨频繁。

## 四 地表覆盖特征

三江源国家生态保护综合试验区土地总面积39.5万平方公里。其中，草地28.02万平方公里，占70.94%；林地2.23万平方公里，占5.65%；耕地1178平方公里，仅占0.30%；其他农用地431平方公里，占0.11%；城镇村庄及工矿用地301平方公里，占0.08%；水域及水利设施用地0.50万平方公里，占1.27%；交通运输用地127平方公里，占0.03%；未利用土地和其他用地8.54万平方公里（包括宜林地），占21.62%。另外，用遥感反演数据得到源区内裸地2.28万平方公里，占5.95%；冰川和永久积雪3073平方公里，占0.80%；因其气候条件特殊，地面寒冻分化作用强烈，土壤发育过程缓慢，成土作用时间短，土壤比较年轻，质地粗，沙砾性强，其组成以细沙、岩屑、碎石和砾石为主。区内土壤大多保水性能差、肥力较低，并容易受侵蚀而造成水土流失。

土壤类型可分为15个土类、29个亚类。因高海拔山地多、海拔差异大、相对高差大，土壤类型由高到低呈垂直地带性分布，从高到低主要有高山寒漠土、高山草甸土、高山草原土、山地草甸土、灰褐土、栗钙土和山地森林土，其中以高山草甸土为主，海拔为3500～4800米，沼泽化草甸土也较为发育，冻土面积较大。

## 五 生物群落

三江源区独特的高寒环境形成了独特的高原生物群落,孕育了大量高原独有的生物物种,是珍贵的种质资源和高原生物基因库。

三江源区的维管束植物现有 87 个科、474 属、2308 种,约占全国植物种类的 8%;其中种子植物种数占全国相应种数的 8.5%;草本植物 422 属、占 89%,2125 种、占 92.07%;灌木 41 属、占 8.7%,144 种、占 6.24%;乔木 11 属、占总属数的 2.3%,39 种、占 1.69%。植物类型以草本植物居多,植被类型主要有针叶林、阔叶林、针阔混交林、灌木丛、草甸、草原、沼泽及水生植被、垫状植被和稀疏植被等 9 个植被类型,分为 14 个群系纲、50 个群系,保护区内有国家二级保护植物油麦吊云杉、红花绿绒蒿、虫草 3 种,列入国际贸易公约附录Ⅱ的兰科植物 31 种,青海省级重点保护植物 34 种。在适应高寒生态环境的演变过程中,一批具有高原特色的农作物、经济果木、中药材等得以进化培育。

三江源区的野生动物资源十分丰富,种类繁多,区系复杂。按我国动物地理区划,三江源区属"青海藏南亚区",动物分布型属"高地型",以青藏类为主,并有少量中亚型以及广布种分布。据调查,区内有兽类 8 目、20 科、85 种,占全国兽类的 16.8%;鸟类 16 目、41 科、238 种(包含亚种 263 种),占全国鸟类的 19%;两栖爬行类 7 目、13 科、48 种;鱼类 6 科、40 种,40% 以上的种类是中国特有物种。国家重点保护动物有 69 种,其中国家一级重点保护动物有藏羚羊、野牦牛、雪豹等 17 种,国家二级重点保护动物有盘羊、藏原羚等 52 种,还有省级保护动物艾虎、沙狐等 32 种。另外,特有的农牧业资源和野生生物资源，为特色农畜产品加工、中藏药业，以及野生动物驯化养殖业创造了良好的基础条件。

## 六　社会经济发展

2015年，三江源区内总人口132.81万，占青海全省总人口的23.1%。其中非农居民人口27.98万，占总人口的21.1%；农牧业居民人口104.83万，占总人口的78.9%。区内居住有汉族、藏族、蒙古族、回族等多个民族，以藏族为主的少数民族人口约占总人口的80.0%。

三江源区虽然人烟稀少，人口密度不到4人/平方公里，但随着人口不断增长，人类活动对该地区的生态环境保护形成了较大压力。据统计，从清朝中期到民国初年的100年间，三江源区的人口增长75.0%，年均增长率为5.6‰，平均125年翻一番；民国时期总人口增长了90.0%，年均增长率为18.6‰，平均38年翻一番。新中国成立后到2010年第六次人口普查时，人口总规模增长了2.48倍，年均增长率高达20.6‰，平均21年翻一番。

三江源区的传统经济以草地畜牧业为主，是一个社会经济基础薄弱、生产方式相对落后的地区。新中国成立前，三江源的工业生产几乎是一片空白，仅有少量的传统手工业和民族用品工业。新中国成立后，建成了煤炭、电力、木材加工、建筑建材、采金、农牧机械修理和农畜产品加工等工业企业。近年来，三江源区又集中打造了一批集生态旅游、探险旅游、宗教朝觐和风情旅游于一体的特色旅游项目，第三产业蓬勃发展。

2015年，三江源区的生产总值309.16亿元，占青海省的12.8%。其中，第一产业82.43亿元，第二产业132.54亿元，第三产业94.19

亿元。公共财政收入22.55亿元，公共财政支出313.02亿元，支出为收入的13.9倍。全年居民人均可支配收入1.24万元，仅为全国人均水平的56.2%。总体而言，三江源区经济发展滞后，地方财政收不抵支，贫困面广、量大、程度深，有8个国家扶贫工作重点县、8个省扶贫工作重点县，贫困人口比例高。

# 分 报 告

Subject Reports

# G.2
# 水文水资源状况

## 一 降水特征

降水是地表水资源的主要来源,其时空分布与变化直接决定了一个地区的干湿程度和水资源量等。三江源地处高原腹地,具有海拔高、地形复杂、蒸发强烈等特点,该地区的降水也表现出不同于其他地区的独特特征。因此,研究三江源区的降水时空分布规律和变化趋势对理解该地区水文水资源状况具有重要意义。

由于该区域人口相对稀少,无人区面积大,近 40 万平方公里的区域范围内只有 23 个长期监测的雨量站点,空间覆盖密度极低,地面监测数据远不能真实反映全区的降水量分布情况。随着卫星遥感技术的不断进步,卫星降水数据产品能够有效弥补实测站点不足的缺陷。目前,国际上使用最为广泛的卫星降水数据产品有两种,一种是美国国家海洋与大气局（NOAA）的CMORPH数据，另一种是美国宇航局（NASA）和日本宇航探索局（JAXA）联合发布的TRMM数据。研究表明，这两套数据在青藏高原地区的表现比较一致，但是CMORPH数据的空间分辨率更高（CMORPH数据最高空间分辨率达到8公里，即每隔8公里就有一个数据点，而TRMM数据空间分辨率只有0.25度，约20公里），因此CMORPH数据在该区域更具优势。CMORPH数据的采集始于1998年，由于卫星最初运行的前两年中数据采集与处理方法尚未成熟，数据精度相对较低，因此最终采用2000～2015年的CMORPH数据对三江源区的降水进行分析。多数科研机构对该数据的研究和使用结果表明，该项技术精度较高，尤其是研究大范围的降雨分布可信性较高。

### （一）空间分布

基于CMORPH卫星降水数据，三江源区2000～2015年多年平均年降水量空间分布如图2.1所示。从整体上来看，三江源区的降水呈现自东南向西北逐渐减少的空间分布特征，降水量在空间上的变化较为均匀。西部和北部广大地区年降水量普遍低于300毫米，属于干

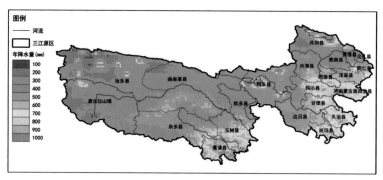

图2.1　2000～2015年三江源区年降水量空间分布示意

旱半干旱地区，中部地区降水量在 300 ~ 700 毫米，东部和南部地区的年降水量可达 700 毫米以上，属于半湿润地区。2000 ~ 2015 年三江源区多年平均年降水量为 406.6 毫米，年均降水资源总量约为 1606 亿立方米，其中长江流域和黄河流域所占比例十分接近，均为 36.5%，澜沧江流域和内流区的降水资源量各占 14.1% 和 12.8%。

需要说明的是，CMORPH 卫星降水数据对内陆水体的降水反演存在系统偏差，导致对三江源湖泊地区的降水量估计明显偏大。从图 2.1 可以看出，在青海湖、鄂陵湖、扎陵湖和羌塘高原内陆湖区，降水量在空间上骤然增高，反常地高于湖泊以外的地区。这一问题在对其他地区监测时也同样出现，例如，国外学者曾指出 CMORPH 和 TRMM 卫星降水数据在美国东北地区的内陆湖区存在显著的系统性高估。尽管卫星降水数据的这一问题会增大三江源区降水量估计的误差，但对我们研究降水的空间分布规律影响不大。

长江流域在三江源区内占地面积 15.8 万平方公里，除干流金沙江发源于三江源区外，长江支流岷江和雅砻江同样发源于三江源区。年均降水量为 366.1 毫米，年均降水资源总量为 586.7 亿立方米。金沙江上游流域处于三江源西部降水较少的地区，年降水量在 150 ~ 400 毫米，岷江和雅砻江上游降水量较多，年降水量在 500 ~ 750 毫米。

黄河流域在三江源区内占地面积 11.9 万平方公里，年均降水量为 487.3 毫米，年均降水资源总量为 586.6 亿立方米。该地区降水量分布东高西低，东部的同仁县、泽库县、同德县、玛沁县、河南蒙古族自治县、甘德县、久治县年降水量较多，可达 500 毫米，甘德县以上的黄河流域以及龙羊峡水库一带，年降水量普遍低于 500 毫米。

澜沧江流域在三江源区内占地面积 3.7 万平方公里，年均降水量为 605.0 毫米，年均降水资源总量为 226.5 亿立方米。降水量沿河流从 300

毫米增加到800毫米以上，处于三江源区内降水较为丰沛的地区。

其他区域属于内流区，共占面积8.1万平方公里，虽然内流区的降水不直接汇入三江，但是对当地生态系统仍有重要影响。内流区年均降水量为268.3毫米，年均降水资源总量为206.4亿立方米。

### （二）时间分布

三江源区的降水具有明显的季节性，2000～2015年各月平均降水量变化过程如图2.2所示。5～9月的降水量占全年降水量的85.5%，其中6～8月降水量就占了全年的59.6%，7月的降水是全年的最高值。降水量时间分布不均的特性加重了三江源区的干旱形势，给水资源的管理带来了挑战。

2000～2015年三江源区内各流域的降水量随月份变化过程如图2.3所示，可以看出各流域降水的季节性都十分明显。其中澜沧江流域的降水量最高，黄河流域次之，再次为长江流域与内流区，两者降水量较为接近。考虑到各流域在三江源区内的面积不同，降水量的差异并不代表各流域降水资源总量的差异。

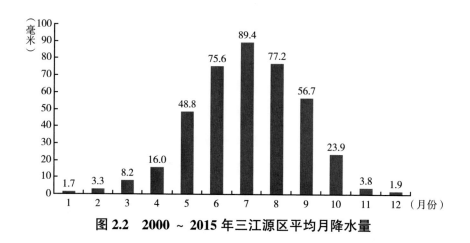

图2.2　2000～2015年三江源区平均月降水量

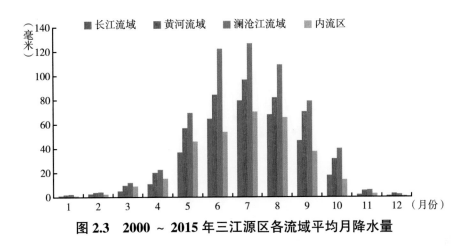

图 2.3　2000～2015 年三江源区各流域平均月降水量

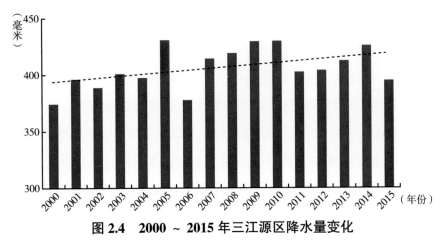

图 2.4　2000～2015 年三江源区降水量变化

2000～2015 年三江源区降水量变化过程如图 2.4 所示，年降水量在 370～430 毫米波动，变幅不大。数据分析表明，三江源区年降水量上升趋势显著，增幅约每年 1.69 毫米，趋势线如图 2.4 中虚线所示。

由于卫星降水的观测年限较短，为了观察三江源区降水量的长期变化，科考队使用雨量站资料延长卫星降水序列。用泰森多边形法将三江源区内 23 个雨量站的降水数据处理成全区平均值，得到 1960～2015 年三江源区降水量变化图（见图 2.5）。结果表明，该地区

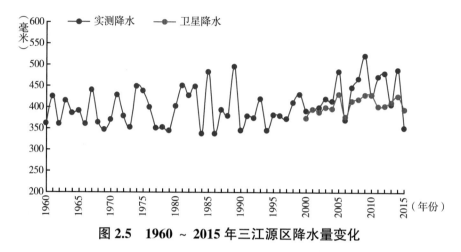

图 2.5　1960～2015 年三江源区降水量变化

平均年降水量在 400 毫米左右波动，1994 年之前降水量变化趋势不明显，1994 年之后降水量开始上升，2000～2015 年的平均降水量为 434.3 毫米，比卫星降水给出的 406.6 毫米稍高，且大部分年份的实测降水都高于卫星降水。由于三江源区气象站分布稀疏，空间精细度不足，其在研究降水的空间特性时与卫星降水数据相比准确度不足，仅作为参考。

基于卫星降水的各流域降水资源量年际变化如图 2.6 所示。可以看到，各流域降水丰枯变化规律并不同步，变化趋势各自较为独立。

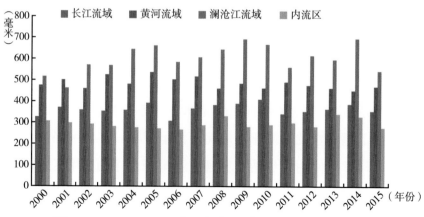

图 2.6　2000～2015 年三江源区各流域降水总量变化

### （三）卫星数据验证

为了验证卫星数据的可靠性，科考队将雨量站实测数据与之对比。雨量站实测资料相对可靠，因此被视为降水量真值。三江源区内 23 个雨量站的分布如图 2.7 所示，其中大部分站点都集中在三江源区的东部和中部。将雨量站逐年降水量与所处区域的卫星数据进行对比分析，发现除在少数几个站点误差较高外，卫星数据的年降水量的平均误差在 10%～20%。整体而言，所处地区的海拔越高，卫星降水数据的精度就越低。但是，由于西部高海拔地区的降水量较低，对全区的降水总量贡献不大，因此误差的影响相对较小。从图 2.8 可以看出，除沱沱河站的误差较高（25%）外，卫星数据在曲麻莱站、达日站、兴海站与实测数据的符合度较高，相对误差较小。沱沱河站所处位置海拔高、地形复杂，在站点附近 20 公里验证范围内地形变化剧烈，不具备代表性，而其他三个气象站所处位置相对平缓，数据相对可靠。

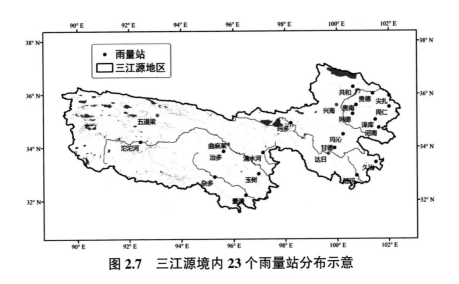

图 2.7　三江源境内 23 个雨量站分布示意

相对于雨量站实测数据，卫星降水数据在空间覆盖度上具备明显的优势，尤其对于雨量站稀少的地区，卫星反演降水能够在更大的范围宏观上监测降水的时空分布特征，帮助发现新的规律，这也是本文采用卫星数据进行降水分析的原因。

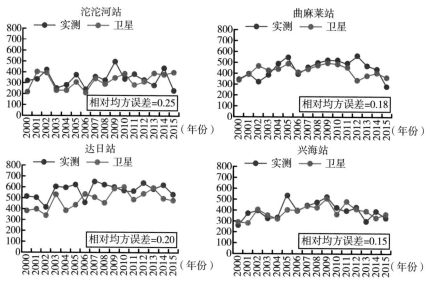

图2.8　气象站实测降水与卫星降水数据对比

## 二　河流水系

三江源区水系可以分为外流水系和内流水系。外流水系包括长江、黄河、澜沧江三条大河的源头流域，内流水系包括羌塘高原水系、柴达木盆地水系和青海湖水系。基于清华大学自主研究和提取的高分辨率数字河网"清华Hydro30"，对三江源区的河流水系结构进行研究。"清华Hydro30"基于30米的全球DEM数据，采用改进的河网提取算法提取而成，具有河网精细程度高、位置准确、级别丰富等优点。三江源区地形复杂,河网密布,存在大面积的无人区,科考队难以对所有河流进行实地测量,而"清华Hydro30"技术能够准确识别该地区河流水系的分布和结构,对了解掌握三江源区河流状况并进一步实施生态保护措施具有重要意义。流域分布情况如图2.9所示。三江源区外流区三大流域的面积与干流长度和支流数目如表2.1所示,经统计,流域面积在50平方公里以上的河流有1416条。

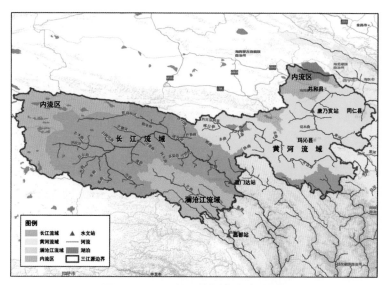

图 2.9　三江源区流域分布示意

### 表 2.1　三江源区内三大流域的基本信息

（a）流域面积与干流长度

| 流域 | 面积（万平方公里） | 占三江源区总面积的比例（%） | 干流长度（公里） | 占干流全长的比例（%） |
|---|---|---|---|---|
| 黄河流域 | 11.9 | 30.0 | 1983 | 34.9 |
| 长江流域 | 15.8 | 40.0 | 1206 | 19.1 |
| 澜沧江流域 | 3.7 | 9.4 | 448 | 20.4 |

**（b）各级别支流数目**

单位：条

| 流域 | 支流数目 | | | |
|---|---|---|---|---|
| | 一级支流 | 二级支流 | 三级支流 | 四级支流 |
| 黄河流域 | 126 | 338 | 157 | 8 |
| 长江流域 | 109 | 274 | 162 | 30 |
| 澜沧江流域 | 46 | 108 | 51 | 7 |

注：最小流域面积设为50平方公里。

## （一）黄河源区水系

黄河发源于巴颜喀拉山北麓的约古宗列盆地隅，源头海拔4724米，在寺沟峡流入甘肃，大体呈"S"形走势。黄河自河源至三江源出口处全长1983公里，占干流全长5687公里的34.9%。黄河流域在三江源区内的面积为11.9万平方公里，占三江源区总面积的30.1%，其中流域面积大于50平方公里的黄河一级支流有126条，其中流域面积大于5000平方公里的有4条（切木曲、多曲、热曲、曲什安河），1000～5000平方公里的有21条，500～1000平方公里的有45条。二级及二级以下支流众多，黄河源区主要支流如图2.10所示。黄河支流中位于三江源区且流域面积较大、径流量较大的流域包括多曲流域、热曲流域、柯曲流域、达日河流域、西科曲及东科曲流域、切木曲流域、曲什安河流域、大河坝河流域、巴曲流域、茫曲流域、隆务河流域等。

多曲发源于称多县北部，位于扎陵湖南岸，流入玛多县鄂陵湖，流域面积为5905.14平方公里，海拔4400～4600米（来源于谷歌地球，下同），由西南向东北递减。多曲流域属寒温带大陆性季风气候，年平均降雨量为300～400毫米（来源于水文年鉴，下同）。多年平均年径流量4.2亿立方米，自然落差536米（出自《中国水系辞典》，

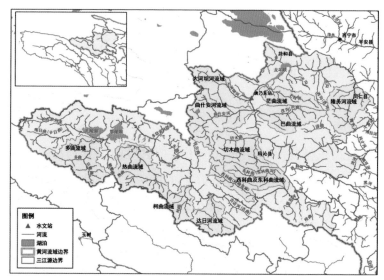

**图 2.10 黄河源区主要支流流域示意**

下同)。多曲中下游河道两侧多沼泽及草地,牧业发达。

热曲发源于玛多县南部、达日县北部,主要有东部的热曲和西部的黑河两条支流,热曲和黑河汇合后流入黄河。流域大部分属于玛多县,海拔 4200～4800 米,地势西南高东北低,流域面积为 6702.00 平方公里。热曲流域属高寒草原气候,年平均降雨量为 400～500 毫米。热曲多年平均年径流量 6.8 亿立方米,自然落差 585 米。热曲流域内草地分布广泛,水草丰盛,经济以畜牧业为主。

柯曲发源于青海省达日县与四川省石渠县分水岭,自南向北流入黄河,流域面积为 2455.91 平方公里,海拔 4300～4900 米。柯曲流域属高寒半湿润气候,无明显四季之分,年平均降雨量为 500～600 毫米。该流域草地覆盖面积广,占 90% 以上。流域所在的达日县是果洛州重要的商品、汉藏药材集散地和交通枢纽。

达日勒曲发源于达日县中部及班玛县西北部,大部分位于达日

县,主要有达日勒曲和都曲两条支流,汇合后自南向北汇入黄河,流域面积为3400.61平方公里,海拔4100~4800米。达日勒曲流域属高寒半湿润气候,年平均降雨量为600~700毫米。达日河多年平均年径流量4.9亿立方米,自然落差116米。流域内民族主要为藏族,还有少数汉族、蒙古族、回族。

西科曲及东科曲流域相邻,位于甘德县境内,分别自西北到东南汇入黄河,其中西科曲流域面积为2584.38平方公里,东科曲流域面积为3477.37平方公里,海拔3800~4600米。西科曲及东科曲流域属高原大陆性半温润气候,年平均降雨量为600~700毫米。流域内民族以藏族为主,有极少数汉族、回族和土家族。

切木曲发源于玛沁县转山-玛卿岗日山,自西向东汇入黄河,流域面积为5610.32平方公里,海拔3700~5400米,转山-玛卿岗日山主峰海拔6282米。切木曲流域属于大陆性寒润性气候,年平均降雨量为500~600毫米。切木曲多年平均年径流量2.5亿立方米/秒,自然落差1830米。玛沁县草地资源丰富,牧业发达。

曲什安河流域发源于玛多、玛沁、兴海三县接壤处,流经兴海县自西向东汇入黄河,流域面积为6593.72平方公里,地势西南高东低,海拔3100~5200米。曲什安河流域属于高原大陆性气候,年平均降雨量300~400毫米。曲什安河大米滩水文站多年平均年径流量8.1亿立方米,年平均水面蒸发量692.2毫米。曲什安河上游为峡谷,中下游为草地。该流域是全省冬虫夏草的主产区之一。

大河坝河流域发源于兴海县西部,自西北向东南汇入黄河,流域面积为4028.89平方公里,海拔3000~4700米。大河坝河流域属高原大陆性气候,年平均降雨量300~400毫米。大河坝河上村水文站多年平均径流量3.6亿立方米。大河坝河两岸为草地及荒漠地域,水源不丰。

巴曲流域发源于泽库县西部、同德县东部，主要有巴沟和尕曲两条支流，汇合后自西向东汇入黄河。巴曲流域面积为 4286.78 平方公里，海拔 2800～4200 米。巴曲流域属于高原大陆性气候，年平均降雨量为 400～500 毫米。巴曲多年平均年径流量 3.2 亿立方米，自然落差 1208 米；上游流经高原草地，水源不丰，中游河段为季节性河流。流域内已修建水电站一座，装机容量 1600 千瓦。

茫曲流域发源于贵南县东部，自西向东汇入黄河龙羊峡水库，流域面积为 2997.14 平方公里，海拔 2800～3800 米。茫曲流域属于高原大陆性气候，年平均降雨量为 300～400 毫米。茫曲多年平均年径流量 1.5 亿立方米，自然落差 1524 米。流域内已建发电站两座，总装机容量 1800 千瓦。该流域是贵南县重要的农业生产基地。

隆务河流域发源于泽库县西部，自南向北流经同仁县汇入黄河，流域面积为 5035.40 平方公里，地势西高东低，海拔 2800～4000 米。隆务河流域属于高原大陆性气候，年平均降雨量 300～500 毫米。隆务河口水文站多年平均年径流量 6.1 亿立方米，自然落差 2292 米。流域内农牧业占主导地位，农作物以小麦、青稞为主。

## （二）长江源区水系

长江正源沱沱河发源于唐古拉山中段的各拉丹东雪峰，它与南源当曲在囊极巴陇汇合后称通天河，继而与北源楚玛尔河相汇，向东南流至玉树县接纳巴塘河后称通天河。长江流域在三江源区内的流域面积为 15.8 万平方公里，占三江源区总面积的 40%，干流长 1206 公里，占干流总长的 19.1%，落差 2065 米，平均比降 1.78‰。三江源区内流域面积大于 50 平方公里的长江一级支流有 109 条，其中流域面积大于10000平方公里的有4条（当曲、楚玛尔河、岷江、雅砻江），5000～10000平方公里的有7条，1000～5000平方公里的有22条，500～1000平方公里的有28条，长江源区主要支流如图2.11所示。长江支流中位于三江源区且流域面积较大、径流量较大的流域包括当曲流域、楚玛尔河流域、扎木曲流域、莫曲流域、北麓河流域、科欠曲流域、色吾曲流域、聂恰曲流域、德曲流域、雅砻江流域、岷江上游流域等。

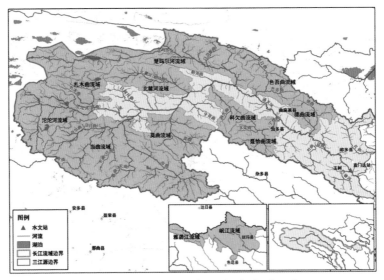

图2.11　长江源区主要支流流域示意

通天河为长江上游干流之一段，特指沱沱河与当曲汇合处至玉树县巴塘河汇入口的长江干流段。位于青海省西南部的玉树藏族自治州境内，流经曲麻莱、治多、称多和玉树四县。因地处"世界屋脊"青藏高原，地势高峻，曾传为通天之河，故称通天河。通天河自正源沱沱河与南源当曲汇口开始，向东流进入治多县，至莫曲汇口折向东北流，在北麓河口附近成为治多、曲麻莱两县界河，至科欠曲汇口以下不远改向东南流，在得列楚拉勃登，长江北源楚玛尔河汇入，成为汹涌澎湃的大河，自此流出长江源区；继续向东南流，在德曲汇口以下成为玉树、称多两县界河，至玉树县形成附近的巴塘河汇口为止，以下即金沙江。通天河干流呈弓形，全长 813 公里，河口控制流域面积近 14 万平方公里，年平均流量 400 立方米/秒，年径流量约 130 亿立方米，输沙量 900 多万吨、含沙量 0.74 千克/立方米，河水清澈，水质良好。

2016 年 6 月，三江源国家公园综合科学考察队赴三江源区进行实地水文勘察，共计对 10 条河流的 12 个断面进行了流量测验及水文要素测量。测流断面所在位置如图 2.12 所示。

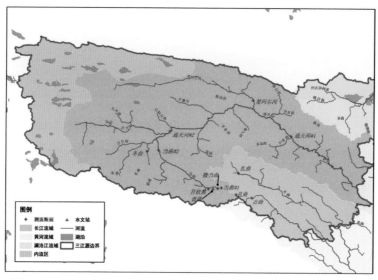

**图 2.12　三江源国家公园综合科学考察的测流断面位置示意**

科考队在通天河选取了两个断面进行测量。通天河 1 号测流断面位于治多县国道 309 线通天河大桥处，地理坐标为东经 95°49′25″、北纬 34°02′12″，断面所在河道为砂砾石河床，大桥上游水流较为集中，在大桥下游水流开始分散，形成两股水流，左岸为陡坎，右岸为平缓砂砾石河滩，周边地势平坦开阔，植被良好。流量测验采用ADCP进行施测，考察当日（2016年6月24日），测验断面水面宽224米，断面面积225平方米，流量256立方米/秒（见图2.13）。

（a）通天河大桥测流断面实景

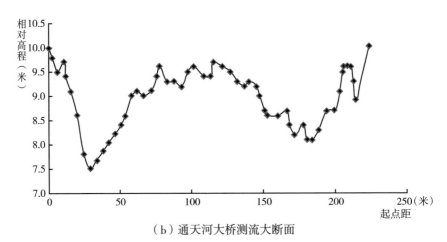

（b）通天河大桥测流大断面

图2.13　通天河大桥测流断面

通天河2号测流断面位于治多县囊极巴陇公路桥处，地理坐标为东经93°01′27″、北纬34°08′56″，海拔4463米。断面所在河道为砂砾石河床，水流较为集中，上游顺直河道长约220米，下游顺直河道长约410米，桥面到水面高7米，两岸均为平缓砂砾石河滩，周边地势平坦开阔，植被良好。流量测验采用ADCP进行施测。考察当日（2016年6月25日），测验断面水面宽165米，断面面积213平方米，流量260立方米/秒（见图2.14）。

（a）通天河囊极巴陇河段实景及测流断面

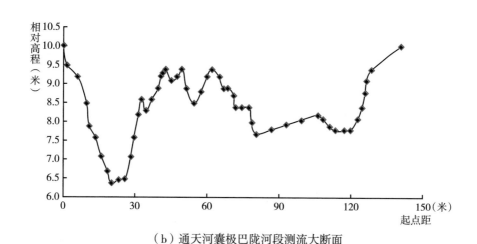

（b）通天河囊极巴陇河段测流大断面

图2.14 通天河囊极巴陇河段测流断面

当曲又名阿克达木河，是长江南源，位于青海省西南隅，发源于杂多县境唐古拉山脉东段北支霞舍日阿巴山，流经玉树州杂多县、格尔木市唐古拉山镇，在治多县境内囊极巴陇处与正源沱沱河汇合，汇口以下称通天河。当曲全长352公里，流域面积为31269.43平方公里，海拔4600～5100米，年平均流量146立方米/秒，年径流量46.02亿立方米。当曲流域属于大陆性季风气候，受地理环境的影响，全年气候寒冷，无四季之分，年平均降水量700～900毫米。流域内民族以藏族为主，还有汉族、回族、蒙古族等。流域内为天然牧场，水草丰盛，第一产业从业人数占所有产业从业人数的90%以上；在农业和牧业中，又以牧业为主。

三江源国家公园综合科学考察队在当曲选取了两个断面进行测量。当曲1号测流断面位于杂多县查旦乡公路桥上游105米处，地理坐标为东经94°12′54″、北纬32°52′11″，海拔4740米，周围地势平缓开阔，为山地草原，植被良好。断面所在河道为砂砾石河床，水流平稳、集中，左岸为草地，右岸为砂砾石河滩。流量测验采用悬杆流速仪进行施测，考察当日（2016年6月22日），测验断面水面宽8.9米，断面面积2.15平方米，最大水深0.36米，最大流速0.35米/秒，流量0.581立方米/秒（见图2.15）。

（a）当曲查旦乡公路桥上游段实景及测流断面

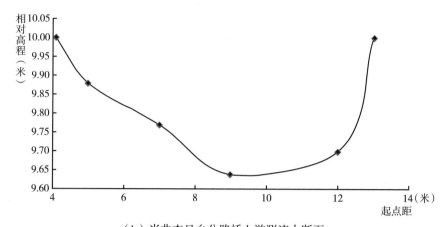

（b）当曲查旦乡公路桥上游测流大断面

**图 2.15　当曲查旦乡公路桥上游测流断面**

当曲 2 号测流断面位于治多县索加乡尕日曲汇合口上游 30 公里公路大桥处，地理坐标为东经 92°45′34″、北纬 33°42′40″，距离国道 109 线雁石坪 75 公里。断面所在位置海拔 4510 米，集水面积 1.61 万平方公里，至河口囊极巴陇 61 公里，所在河道为砂砾石河床，河道顺直，水流平稳、集中。流量测验采用 ADCP 进行施测。考察当日（2016 年 6 月 26 日），测验断面水面宽 80 米，断面面积 86.4 平方米，平均流速 1.28 米/秒，流量 111 立方米/秒（见图 2.16）。

（a）当曲公路大桥处实景及测流断面

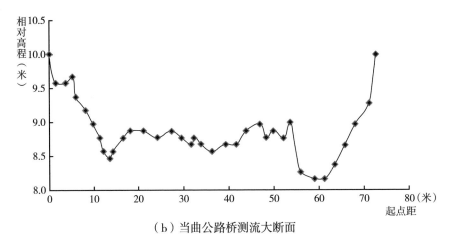

（b）当曲公路桥测流大断面

**图 2.16　当曲公路桥测流断面**

楚玛尔河为长江北源，发源于昆仑山脉南支可可西里山东麓，在治多县境西部。源头有两支：北支发源于可可西里湖东南约18公里处，源头海拔4920米，全长46公里；西支发源于可可西里湖南侧的黑脊山南麓，峰顶海拔5432米，全长45.7公里。两支源流汇合后，东流94公里注入叶鲁苏湖（又名多尔改错或错仁德加湖），干流穿湖20公里流出，又流经117公里至楚玛尔河沿青藏公路桥，进入曲麻莱县境内，桥下年平均流量7.78立方米/秒。水流继续迂回蜿蜒东流，下游逐渐南转，入曲麻滩；经曲麻河乡政府驻地南下，在莱涌滩汇入通天河，河口海拔4216米。楚玛尔河全长515公里，流域面积2.08万平方公里。年平均流量约33.0立方米/秒，年径流量9.28亿立方米。楚玛尔河流域地势西高东低，海拔4600～5200米，属于高原大陆性气候，年平均降雨量200～300毫米。流域上游地区属于可可西里的一部分，是著名的无人区，区域内各种国家级保护动物——藏羚羊、藏野驴、野牦牛等分布广泛。

三江源国家公园综合科学考察队在楚玛尔河选取位于曲麻莱县曲麻河乡曲麻河大桥处的河流断面进行测量，地理坐标为东经

94°56′32″、北纬34°51′19″。断面所在河段左岸为冲积平原，河流右岸接近山体，水流沿山根流淌，附近河段水流呈网状分布，水流较分散。该河段共建有上、中、下三座跨河大桥，断面所在位置海拔4270米，在一代桥附近水流相对比较集中，中低水位时，左岸桥下干枯，水流主要集中在右岸桥下通过，水面宽70～90米，主流距离桥右岸桥头26～50米。流量测验采用ADCP进行施测。考察当日（2016年6月24日），测验断面水面宽41米，断面面积9.8平方米，流量5.42立方米/秒（见图2.17）。

（a）楚玛尔河曲麻河大桥段实景及测流断面

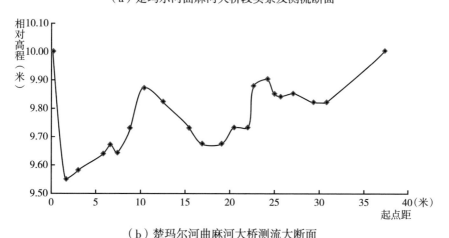

（b）楚玛尔河曲麻河大桥测流大断面

图2.17 楚玛尔河曲麻河大桥测流断面

扎木曲发源于唐古拉山镇及治多县边境，自北向南汇入长江（通天河），流域面积为5293.98平方公里，海拔4600～5000米。扎木曲流域属于温带半湿润高原季风气候，冬冷夏凉，雨水充沛，日照相对较少，年平均降雨量800～900毫米。扎木曲多年平均年径流量2.4亿立方米，自然落差647米。

莫曲流域与当曲流域毗邻，莫曲发源于杂多县北部那日根山，自南向北流经治多县汇入长江（通天河）。流域面积为8892.04平方公里，地势南高北低，海拔4500～5000米。莫曲流域属于高原大陆性气候，年平均降雨量700～800毫米。莫曲多年平均年径流量6.9亿立方米，自然落差610米。

北麓河发源于曲麻莱县西部贡昌日玛西南麓，自西向东汇入长江（通天河）。流域内河网密集，流域面积为7962.82平方公里，海拔4300～5000米。北麓河流域属于高原高寒气候，年平均降雨量800～900毫米。北麓河自然落差613米，多年平均年径流量6.6亿立方米。北麓河上段称日阿池曲，为季节性河流；下段河流常年有水。北麓河流经的大部分地区为沙地戈壁滩，人烟稀少。

科欠曲发源于治多县中部，自南向北汇入长江（通天河），流域面积为3567.72平方公里，海拔4300～5200米。科欠曲流域属于高原大陆性气候，年平均降雨量600～800毫米。科欠曲自然落差1085米，多年平均年径流量2.8亿立方米。流域内民族以藏族为主，还有汉族、回族、蒙古族等；流域内经济以牧业为主。

色吾曲发源于曲麻莱县中部，北与黄河的源头毗邻，主要有昂日曲和色吾曲两条支流，汇合后自东北向西南汇入长江。流域面积为6753.69平方公里，海拔4500～4800米。色吾曲流域属于高原高寒气候，年平均降雨量700～800毫米。色吾曲自然落差536米，多年平均年径流量5.4亿立方米。流域内多深山峡谷，植被覆盖良好，牧业发达。

聂恰曲发源于杂多县与治多县交界处的曲阿加吉玛峰（海拔5930米），自西南向东北汇入长江。流域面积为5695.29平方公里，海拔4400～6000米。聂恰曲流域属于高原大陆性气候，年平均降雨量550～650毫米。聂恰曲流经治多县县城，当地居民以藏族为主，还有汉族、回族、蒙古族等；聂恰曲下游筑有引水式发电站一座。

德曲发源于曲麻莱县与称多县边界，自北向南汇入长江。流域面积为4234.60平方公里，海拔4300～5000米。德曲流域属于寒温带大陆性季风气候，年降雨量550～650毫米。德曲自然落差924米，平均年径流量9.4亿立方米。境内多沼泽，水流时断时续。流域内植被覆盖良好，牧业发达。

雅砻江的干流和支流鲜水河均发源于三江源区，干流发源于称多县巴颜喀拉山南麓，支流鲜水河发源于达日县巴颜喀拉山南麓，在三江源区内总面积约1.2万平方公里，海拔4200～4900米。

岷江属于长江中游的一条支流，发源于三江源区。岷江流域在三江源区域分布于达日县、久治县、班玛县境内，面积约9775平方公里，海拔3600～4400米。流域内河网众多，包括大渡河、杜柯河、马儿曲、阿柯河等河流。

（三）澜沧江源区水系

澜沧江发源于唐古拉山北麓的查加日玛西南部，河源海拔5388米，其干流在青海境内称扎曲，由囊谦县境内流入西藏。三江源区内流域面积3.7万平方公里，占三江源区总面积的9.4%，河长448公里，占国境内干流总长的20.4%，落差1553米，平均比降3.47‰。澜沧江源区水系发育，呈树枝状。除干流扎曲外，昂曲和子曲也发源于三江源区境内（见图2.18）。

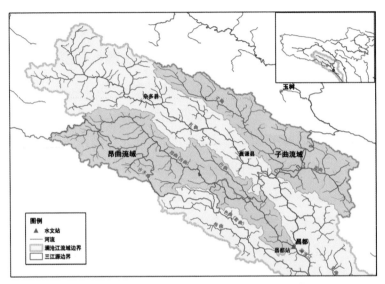

**图2.18　澜沧江上游主要支流流域示意**

扎曲又叫作杂曲，发源于玉树藏族自治州杂多县西北，唐古拉山北麓的查加日玛西4公里的高地，河源海拔5388米。干流自河源至陇冒曲汇口名加果空桑贡玛曲，以下至尕纳松多名扎纳曲，再下始称扎曲。河水自西北流向东南，流经囊谦县境后出省进入西藏自治区。青海省境内干流河长448公里，青海与西藏共界河长约4.5公里，流域面积1.85万平方公里，年平均流量138立方米/秒，多年平均年径流量43.52亿立方米。

三江源国家公园综合科学考察队在扎曲选取位于杂多县扎那曲与扎阿曲交汇处扎阿曲大桥的河流断面进行测量，地理坐标为东经94°36′44″、北纬33°11′58″，海拔4370米。断面所在河道为砂砾石河床，桥下水流分散，在桥下游60米处重新汇合，右岸为平缓砂砾石河滩，左岸为草地，周边地势平坦开阔，植被良好。流量测验采用ADCP进行施测。考察当日（2016年6月23日），测验断面水面宽47米，断面面积42.7平方米，流量61.2立方米/秒（见图2.19）。

（a）扎曲实景及扎阿曲大桥测流断面

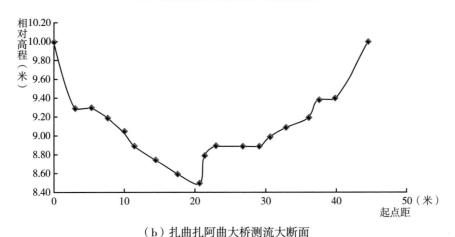

（b）扎曲扎阿曲大桥测流大断面

**图2.19 扎曲扎阿曲大桥测流断面**

昂曲发源于西藏巴青县北部、青海杂多县西部，横跨杂多县、囊谦县、类乌齐县，在昌都县汇入澜沧江。流域面积为1.68万平方公里，地势西高东低，海拔3600~5200米。昂曲多年平均流量186立方米/秒，自然落差 1898 米。昌都是藏东经济中心，交通方便，物资文化交流频繁。

子曲发源于杂多县东南部山丘，流经囊谦县，在西藏境内汇入澜沧江，流域面积为 1.29 万平方公里。子曲多年平均流量 137 立方米/秒，自然落差 1540 米。子曲流域水系发达，河网密集，水资源丰富。

## （四）内流区水系

位于三江源区西北的羌塘高原属于我国著名的内流湖区，河网密集，湖泊众多。虽然由于地形的阻隔，羌塘内流河湖与长江、澜沧江等外流河分属于不同的流域，两者之间不存在直接的通路，但是，由于存在地下水交换和大气水热交换，以及生态系统相互耦合，两者之间仍然紧密相连。图 2.20 为羌塘高原区的河流及湖泊分布。

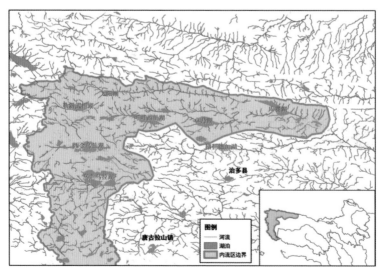

图 2.20　羌塘高原内流湖区的河流及湖泊示意

## （五）河川径流

三江源区河流主要分为外流河和内流河两大类，三条江河多年平均径流量 522.9 亿立方米，其中黄河流出水量 210.6 亿立方米，长江流出水量 186.3 亿立方米，澜沧江流出水量 126.0 亿立方米。雨水和冰雪融水是径流的主要来源。

唐乃亥水文站是黄河上游的重要控制站，控制流域面积 121972 平方公里，控制断面以上河长 1553 公里，占全河长的 28.4%，是龙羊峡的入库站。唐乃亥站 1956～2015 年多年平均年径流量为 201.0 亿立方米，占黄河多年平均天然径流量的 37.7%。径流最大年是 1989 年，为 328.4 亿立方米，最小年是 2002 年，为 106.4 亿立方米，变幅为 222 亿立方米。年径流量在 1983 年后呈减少趋势，2003 年后水量又开始上升。黄河径流年内分布很不均匀，汛期（5～10 月）径流量约占全年径流总量的 78%。黄河源区河川径流主要来自大气降水，其次为冰雪融水和地下水的补给。

自 20 世纪 80 年代以来，黄河入海水量大幅减少。实际上，黄河源区水量并未持续减少。如图 2.21 所示，黄河源区唐乃亥站虽有年际波动，但长期来看基本保持在 200 亿立方米的水平，相对稳定，而黄河入海口花园口水文站的径流量则明显下降。黄河的生命在源头，三江源的生态意义不言而喻，倘若不加以适当的保护，我们的母亲河黄河将遭受严重的生态灾难。

黄河干流流出三江源区的出口是隆务河口，该河口的水文站已被撤除。黄河在三江源区的流出水量，根据河口上游最近的控制站［贵德（二）站］实测径流量，流域面积按比例放大求得黄河干流从三江源区流出的平均年径流量约 210.6 亿立方米。

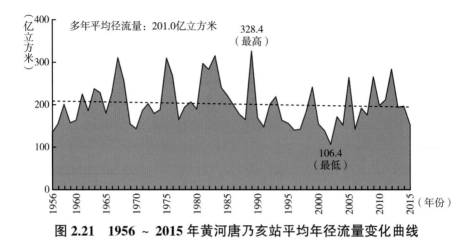

**图 2.21　1956～2015 年黄河唐乃亥站平均年径流量变化曲线**

直门达水文站是长江上游的重要控制站，控制流域面积 137704 平方公里。直门达水文站 1956～2015 年多年平均年径流量为 129.8 亿立方米，占长江多年平均径流总量的 1.3%。径流最大年是 2009 年，为 246.0 亿立方米，最小年是 1979 年，为 70.4 亿立方米，变幅为 175.6 亿立方米（见图 2.22）。年径流量在 1994 年前基本稳定，在 1994 年后径流量增长趋势较为显著。长江径流年内分布很不均匀，汛期（5～10 月）径流量约占全年径流总量的 87%。长江源区径流的主要补给来源，除天然降水外，还有冰川融水。径流的年际变化主要受降水和气候的影响，而年内变化则主要受气温和降水过程的影响，夏秋暖季水量较大。

直门达水文站临近长江干流流出三江源区的出口，流域面积按比例放大得到的多年平均流出水量为 140.0 亿立方米。另外，长江支流雅砻江和岷江也分别贡献了 15.2 亿立方米和 31.1 亿立方米的流出水量。因此，长江从三江源区流出的平均年径流量为 186.3 亿立方米。

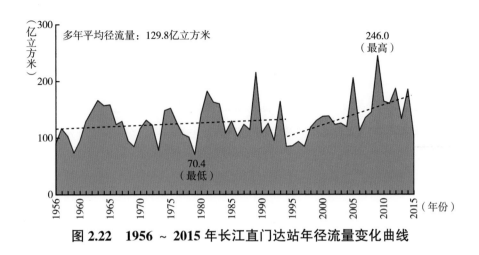

图 2.22　1956～2015 年长江直门达站年径流量变化曲线

根据 2006 年青海省水资源调查成果，澜沧江流出三江源区的多年平均径流量为 126.0 亿立方米，约占国境内流域径流总量的 15%。地表径流比较稳定，水资源丰富。

## 三　湿地资源

三江源区具有丰富的湿地资源。根据 2014 年 6 月 6 日由青海省人民政府新闻办和青海省林业局联合公布的青海省第二次湿地资源调查成果，三江源区湿地总面积为 4.17 万平方公里。其中，长江流域湿地面积为 1.90 万平方公里，占 45.6%；黄河流域湿地面积为 1.22 万平方公里，占 29.3%；澜沧江流域湿地面积为 0.14 万平方公里，占 3.4%；其他内流流域的湿地面积共 0.91 万平方公里，占 21.8%。根据湿地的成因和特点，可分为湖泊湿地、沼泽湿地、河流湿地和人工湿地四大类，三江源区内这四种湿地的面积分别占 21.1%、63.7%、14.2% 和 1.1%。

湖泊湿地是由地面上大小形状不一、充满水体的天然洼地组成的湿地。青海省是一个多湖泊地区，湖泊湿地主要分布在青海湖、可可西里、长江上游、黄河上游以及柴达木盆地南部盐湖地区。湖泊湿地总面积为8775平方公里，其中，长江流域1393平方公里、黄河流域1551平方公里，羌塘高原湖泊群面积达3582平方公里。列入中国重要湿地名录的有扎陵湖、鄂陵湖、玛多湖、黄河源区岗纳格玛错、依然错、多尔改错等。其中扎陵湖、鄂陵湖是黄河干流上最大的两个淡水湖，具有巨大的水量调节功能。

鄂陵湖是黄河流域第一大淡水湖，系断陷构造湖，全部在青海省玛多县境内，湖水面积650.75平方公里，东西宽约34公里，南北长约34.5公里，形如顶角向上的三角形。储水量约116亿立方米，湖的北部水深，南部水浅，平均水深17.81米，最大水深30.7米，位于湖心偏北部，湖底广阔而平坦。湖水澄清，呈青蓝色，水质良好，pH值8.05～8.40。

扎陵湖是黄河流域第二大淡水湖，是由断陷盆地形成的构造湖，与东面的鄂陵湖一山之隔，相距十多公里。扎陵湖水域面积544.73平方公里，东西长约37公里，南北宽约23公里，其形状近似顶角向下的三角形。储水量约49亿立方米，湖北部水深，南部水浅，平均水深9米，最大水深13.1米，位于偏向湖心的东北。湖水清澈，水面呈灰白色，水质良好，pH值8.35～8.50。

表2.2为三江源区主要湖泊的水面面积统计，数据来源于国家基础地理信息中心2017年公布的全球地表覆盖数据（GlobeLand30），统计了2010年各湖泊面积的最大值。

表 2.2　三江源区主要湖泊面积统计（基于 GlobeLand30）

单位：平方公里

| 湖泊 | 面积 | 湖泊 | 面积 |
|---|---|---|---|
| 青海湖 | 4362.96 | 可可西里湖 | 320.97 |
| 鄂陵湖 | 685.15 | 库赛湖 | 284.06 |
| 乌兰乌拉湖 | 591.98 | 卓乃湖 | 270.92 |
| 赤布张错 | 550.35 | 错仁德加 | 249.64 |
| 扎陵湖 | 530.42 | 冬给错纳湖 | 234.25 |
| 西金乌兰湖 | 419.29 | 多尔改错 | 228.86 |
| 龙羊峡水库 | 375.80 | 太阳湖 | 102.23 |

同时，青海省水利厅对青海湖、鄂陵湖等几个主要湖泊进行了实地测量，主要测量数据如表 2.3 所示。基于 GlobeLand30 的湖泊面积统计值与实地测量数据相差不大，精确度较高。

表 2.3　青海省主要湖泊水文特征实地测量成果

| 湖泊 | 平均水位（米） | 平均水深（米） | 最大水深（米） | 面积（平方公里） | 容积（亿立方米） |
|---|---|---|---|---|---|
| 青海湖 | 3193.50 | 18.30 | 26.60 | 4294.00 | 785.20 |
| 鄂陵湖 | 4270.10 | 17.81 | 32.09 | 650.75 | 115.91 |
| 扎陵湖 | 4290.80 | 9.00 | 13.54 | 544.73 | 49.03 |
| 冬给错纳湖 | 4084.90 | 28.78 | 90.22 | 247.13 | 71.12 |

注：扎陵湖包括茶木措、卓让措。

过去近 30 年间，黄河源区和长江源区东南部湖泊数量和面积都表现为大范围萎缩，黄河源区西北部甚至出现部分湖泊消失、湖盆变

为草地的现象,部分湖泊由一个大湖分裂为几个较小的湖泊。相反,在长江源区西部和北部湖泊则表现为扩张态势,部分低洼区域甚至形成了新的湖泊。20世纪70年代末至90年代初,除长江源区西南部以外,三江源区湖泊主要表现为萎缩态势,特别是源区西部。90年代初至2004年,前期萎缩的湖泊开始扩张,特别是长江源区西北部,但是黄河源区的湖泊一直处于持续萎缩的态势。

沼泽湿地具有湿润气候、调节河川径流、补给地下水和维持地区水量平衡的重要作用。三江源区环境严酷,沼泽类型独特。沼泽主要类型有三叶碱毛茛沼泽和杉叶藻沼泽,且大多数为泥炭土沼泽。在黄河源,以及长江的沱沱河、楚玛尔河、当曲河三源头,澜沧江河源都有大片沼泽发育,总面积达2.65万平方公里。

长江流域有沼泽面积1.4万平方公里,沼泽大部分集中于长江源区潮湿的东部和南部,而干旱的西部和北部分布甚少。从地势方面看,沼泽主要分布在河滨湖周一带的低洼地区,尤以河流中上游分布为多,当曲水系中上游和通天河上段以南各支流的中上游一带沼泽连片广布。在唐古拉山北侧,沼泽最高发育到海拔5350米,达到青藏高原的上限,是世界上海拔最高的沼泽。

黄河流域有沼泽面积9037平方公里,沼泽发育受到半干旱气候特征限制,主要分布于河源约古宗列曲、两湖周围及星宿海地区。澜沧江源区大小沼泽总面积为986平方公里,主要集中在干流扎曲段和支流扎阿曲、阿曲(阿涌)上游。其他内河流域有沼泽面积约2500平方公里。

除湖泊湿地和沼泽湿地外,三江源区还有河流湿地面积5898平方公里、人工湿地444平方公里。

## 四 冰川雪山

冰川是我国西部独特的山地景观，冰川的融水补给河流，浇灌着内陆盆地农田，冰川融化过程又会调节气候，降低局地气温。三江源区冰川较多，源区总共有冰川715条，冰川面积约2400平方公里，冰川资源蕴藏量达2000亿立方米。冰川的空间分布如图2.23所示，以长江源区为最多，分布冰川627条，冰川总面积为1247.21平方公里，冰川储量983亿立方米，年消融量约9.89亿立方米；黄河源区有冰川68条，面积达到131.44平方公里，冰川储量可达11.04亿立方米；澜沧江源区冰川在数量上和流域面积上都较小，只有20条，面积为124.12平方公里。

长江源区的现代冰川均属大陆性山地冰川。冰川主要分布在唐古拉山北坡和祖尔肯乌拉山西段，昆仑山也有现代冰川发育。以当曲流域冰川覆盖面积为最大，沱沱河流域次之，楚玛尔河流域最小。雪

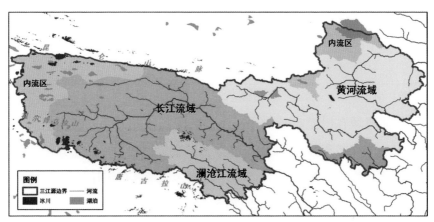

图 2.23 三江源区冰川空间分布示意

山冰川规模以唐古拉山脉的各拉丹冬、尕恰迪如岗及祖尔肯乌拉山的岗钦3座雪山群为大,以各拉丹冬雪山群最为宏伟(见图2.24和表2.4)。

图 2.24　冰川雪山美景(左为昆仑山,右为唐古拉山)

表 2.4　长江源区冰川面积统计

单位:平方公里

| 水系 | 冰川面积 | | |
| --- | --- | --- | --- |
| | 唐古拉山 | 昆仑山 | 小计 |
| 沱沱河 | 380.97 | — | 380.97 |
| 当曲 | 793.4 | — | 793.4 |
| 楚玛尔河 | — | 54.99 | 54.99 |
| 通天河上段 | 12.40 | 5.45 | 17.85 |
| 合计 | 1186.77 | 60.44 | 1247.21 |

黄河流域在巴颜喀拉山中段多曲支流托洛曲源头的托洛岗(海拔5041米),有残存冰川约4平方公里,冰川储量0.8亿立方米,域内有卡里恩卡着玛、玛尼特、日吉、勒那冬则等14座山,海拔5000米以上,终年积雪,多年固态水储量约有1.4亿立方米,两项合计共约2.2亿立方米,年融水量约320万立方米,补给河川径流。

澜沧江源头北部多雪峰，平均海拔 5700 米，最高达 5876 米，终年积雪，雪峰之间是第四纪山岳冰川，东西延续 34 公里长、南北 12 公里宽。面积在 1 平方公里以上的冰川 20 多个。

受气候变暖等因素的影响，三江源区冰川面积在过去的 30 年间整体减少 233 平方公里。观测资料显示，当曲河源冰川退缩率达到每年 9 米时，沱沱河源冰川退缩率达到每年 8.25 米，各拉丹冬的岗加曲巴冰川在近 20 年中后退 500 米，年均后退 25 米。澜沧江源区雪线以下到多年冻土地带的下界，海拔 4500～5000 米，呈冰缘地貌，下部因热量增加，冰丘热融滑塌、热融洼地等类型发育。山北坡较南坡冰舌长 1 倍以上，冰舌从海拔 5800 米雪线沿山谷向下至末端海拔 5000 米左右，最长的冰舌长 4.3 公里。源区最大的冰川是色的日冰川，面积为 17.05 平方公里，是查日曲两条小支流穷日弄、查日弄的补给水源。

## 五　地下水资源

三江源地下水资源蕴藏量比较丰富。根据 2006 年青海省水文水资源勘测局公布的青海省水资源评价报告，三江源区地下水资源总量为 193.3 亿立方米。

长江源区地下水资源量约为 71.2 亿立方米。长江源区地下水属山丘区地下水，主要是基岩裂缝水，其次是松散碎屑岩孔隙水，此外还有冻结层水，其补给来源主要有天然降水的垂直补给和冰雪融水补给，以水平径流为主。地下水分布和降水量分布一致。长江源区普遍分布着地下水上涌所形成的泉涌，河流干支流附近谷地多有密布的泉群，以楚玛尔河下游北岸泉群的泉眼数为最多，分布面积也最广。深

循环的地下水沿断裂通道上涌而形成的温泉在长江源区南北部有出露,以唐古拉山北麓为多,最为集中的温泉群在布曲上段河谷地带。山丘区地下水通过河川外泄,与地表水重合,故长江源区地表水资源量即为水资源总量。

黄河源区地下水资源量约为66.1亿立方米。黄河源区属高原山丘地区,地下水资源动储量,包括山区裂隙水域多年冻土层上部地表活动层潜水,均侧向补给了河川径流而转化为地表水。

澜沧江源区地下水资源量约为45.8亿立方米。澜沧江源区内地下水属山丘区地下水,分布特征主要是基岩裂隙水,其次是碎屑岩孔隙水。补给来源单一,主要接受降水的垂直补给和冰雪融水补给,以水平径流为主,通过河流和潜流排泄。水质较好,pH值7~8.5。

另外,三江源区的内流区蕴藏地下水资源量约10.2亿立方米,其中羌塘高原内流区3.0亿立方米、柴达木盆地1.9亿立方米、青海湖水系5.3亿立方米。

# G.3
# 三江源生态环境保护工作历程及成效

三江源区的生态环境保护和建设在党中央、国务院的高度重视和领导下，各级政府和社会各界都积极行动起来，通过政策、项目、资金、宣传等综合措施，努力加强开展对三江源区生态环境的保护工作，通过十多年的共同保护治理，特别是青海省委、省政府对三江源区的生态环境保护和建设工作采取强有力的措施，取得了显著成效，为保障我国生态安全、促进生态文明建设做出了巨大贡献。

## 一 三江源综合试验区和国家公园的设立

建立"三江源自然保护区"是国家林业局和青海省积极响应国家"再造山川秀美的大西北"的伟大号召和贯彻落实党中央、国务院提出的西部大开发战略的一个重要举措。青海省人民政府于2000年5月批准建立三江源省级自然保护区，保护区总面积达15.23万平方公里，占青海省总面积的21%，占三江源区总面积的42%，涉及果洛藏族自治州6县、玉树藏族自治州6县、海南藏族自治州2县、黄南藏族自治州2县、格尔木市管辖的唐古拉山乡共16县1乡。行政区划上共由69个不完整的乡镇组成。

为早日将三江源自然保护区建设成具有国际影响力的国家级自然保护区，2001年青海省人民政府决定申报三江源国家级自然保护区，并组织编制了《三江源自然保护区总体规划》。2003年1月,国务院批准(国办发〔2003〕5号)三江源国家级自然保护区。三江源国家级自然保护区是三江源区的一部分,占源区面积的42%,是整个地区生态类型最集中、生态地位最重要、生态体系最完整的区域,保护和治理好保护区的生态环境,是恢复源区生态功能的关键。2005年,国务院批准实施《青海省三江源自然保护区生态保护和建设总体规划》,被称为"新世纪中国生态1号工程"的三江源生态保护和建设工程正式启动,规划总投资75亿元。《规划》建设内容包括三大类22项工程,其中生态保护与建设项目有12项,包括退牧还草工程、退耕还林(草)工程、封山育林、沙漠化土地防治、湿地生态系统保护、黑土滩综合治理、森林草原防火、鼠害防治、水土保持工程、保护区管理设施与能力建设、野生动物保护和湖泊湿地禁渔工程;农牧民生产生活基础设施建设项目有6项,包括生态移民工程、小城镇建设、养畜配套工程、能源建设工程、灌溉饲草料基地建设和人畜饮水工程;生态保护支撑项目有4项,包括人工增雨工程、科研课题及应用推广、生态监测和农牧民培训(见表3.1)。

表3.1 三江源自然保护区生态保护和建设总体规划建设内容

| 序号 | 类别 | 建设内容 |
| --- | --- | --- |
| 一 | 生态保护与建设项目(12项) | 退牧还草工程、退耕还林(草)工程、封山育林、沙漠化土地防治、湿地生态系统保护、黑土滩综合治理、森林草原防火、鼠害防治、水土保持工程、保护区管理设施与能力建设、野生动物保护和湖泊湿地禁渔工程 |
| 二 | 农牧民生产生活基础设施建设项目(6项) | 生态移民工程、小城镇建设、养畜配套工程、能源建设工程、灌溉饲草料基地建设和人畜饮水工程 |
| 三 | 生态保护支撑项目(4项) | 人工增雨工程、科研课题及应用推广、生态监测和农牧民培训 |

截至 2013 年底,《青海省三江源自然保护区生态保护和建设总体规划》建设任务全面完成,生态系统退化趋势得到初步遏制;重点生态建设工程区生态状况好转;生态建设任务的长期性、艰巨性凸显。

2008 年,《国务院关于支持青海等省藏区经济社会发展的若干意见》(国发〔2008〕34 号)提出"适时启动三江源二期工程前期研究工作"和"建立三江源国家生态保护综合试验区"的要求。2011 年 11 月,国务院第 181 次常务会议批准实施《青海三江源国家生态保护综合试验区总体方案》,并要求组织编制生态保护、社会事业、基础设施、城镇、产业和水资源等重点领域专项规划。综合试验区范围在三江源国家级自然保护区的基础上,扩大到整个三江源区域,总面积 39.5 万平方公里,包括玉树藏族自治州、果洛藏族自治州、海南藏族自治州、黄南藏族自治州全部行政区域的 21 个县和格尔木市的唐古拉山镇,共 158 个乡镇、1214 个行政村。为了贯彻落实中央领导关于三江源地区生态保护和建设的指示精神,在中央有关部门的支持和帮助下,青海省政府着手规划三江源地区生态保护和建设工程。

根据国家发改委、中央编办、财政部、国土部、环保部、住建部、水利部、农业部、林业局、旅游局、文物局、海洋局、法制办等 13 个部门联合印发的《建立国家公园体制试点方案》,青海省组织编制了《三江源国家公园体制试点方案》。2015 年 12 月 9 日,习近平总书记主持召开第 19 次中央深化改革领导小组会议,审议通过《三江源国家公园体制试点方案》,2016 年 3 月,中共中央办公厅、国务院办公厅印发了《试点方案》,确立了国家公园体制试点的真正开端。青海省委省政府把三江源国家公园体制试点作为"天字号"改革工程,举全省之力以推进,专门成立了由省委书记和省长担任双组长

的三江源国家公园体制试点领导小组,制定了《关于实施〈三江源国家公园体制试点方案〉的部署意见》,组建了管理机构,召开了动员大会,正式启动三江源国家公园体制试点,拉开了我国第一个真正意义上的国家公园建设的序幕。为确保国家公园建设科学推进,三江源国家公园体制试点领导小组部署开展国家公园规划编制工作。

三江源国家公园包括长江源园区、黄河源园区、澜沧江源园区,面积12.31万平方公里。

青海省开展三江源国家公园体制试点是创新生态保护管理体制机制的重要改革举措,实现"两个统一行使"、改变"九龙治水"局面,是国家公园体制试点的核心任务和重要目标。试点工作将着力构建归属清晰、权责明确、监管有效的生态保护管理体制,彻底改变三江源地区各类保护地分属不同部门和行业管理,政出多门、多头管理、体制不顺、权责不清的弊端,将三江源国家公园建设成为生态文明体制改革示范区,为全国生态文明建设探索积累可复制、可推广的新经验。三江源国家公园体制试点是青海省又一项上升为国家战略的生态文明建设重大举措,这对加快青海省乃至全国生态文明制度建设具有里程碑意义。

## 二 自然保护区总体规划实施后成效明显

自然保护区总体规划实施以来,青海省、三江源区各地各有关部门加强组织领导,优化工程布局,健全各项制度,强化项目管理,注重科技支撑,落实惠民政策,进展顺利。2005~2013年,累计投资85.39亿元(不含2011年后实施的草原生态保护补助奖励资金),其中国家投资65.88亿元,地方配套14.18亿元,群众自筹5.33亿元,占规划投资的114%。为了有效开展三江源自然保护区生态保护与建设项目生态成效监测和评估工作,青海省环保厅与水利厅、农牧厅、林业厅、气象局等单位共同组成三江源生态监测工作组,在青海环保厅的组织协调下,以中国科学院地理科学与资源研究所作为技术牵头单位,综合应用地面观测、遥感监测和模型模拟相结合的技术方法,针对生态工程预期目标和区域生态环境特征,在构建综合评估指标体系和生态本底的基础上,共同承担完成了"青海三江源自然保护区生态保护和建设工程(一期)生态成效综合评估"任务。由中国科学院地理科学与资源研究所和青海三江源生态监测组,对总体规划一期生态保护和建设工程成效开展了连续9年的科学监测与评估。评估表明,经过青海省及中央有关部门9年的艰苦努力,三江源地区"生态系统退化趋势得到初步遏制,重点生态建设工程区生态状况好转"。工程取得的成效突出表现为"五增",即增加了植被覆盖度,增加了水资源量,增加了生物多样性,增加了农牧民收入,增进了社会和谐。规划确定的目标基本实现。生态成效显著,具体如下。

(一)林草植被覆盖度增加

2004~2012年,三江源自然保护区植被覆盖度呈现增加趋势,森林覆盖率由2004年的6.09%提高到了2012年的6.99%;草原植被覆盖度平均提高11.6个百分点。2004~2008年,三江源地区草地退化态势与工程实施前相比,退化状态不变类型面积占原退化总面积的69.35%,轻微好转类型面积占21.87%,明显好转类型面积占7.40%,退化发生类型面积占0.81%,退化加剧类型面积占0.57%。图3.1表明草地退化态势得到初步遏制,部分退化草地趋于好转。

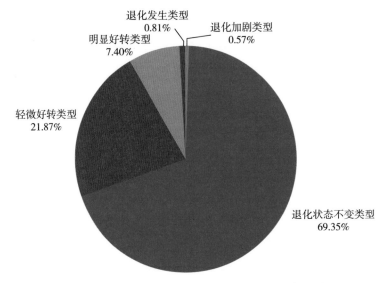

图 3.1 工程实施后三江源地区草地退化类型占比

### （二）江河径流量稳中有增

1975～2004年，长江直门达站出省年平均径流量124.3亿立方米；2004～2012年，出省年平均径流量164.2亿立方米，年平均增加39.9亿立方米；2015年出省径流量为155.02亿立方米。1975～2004年，黄河唐乃亥站出省年平均径流量201.9亿立方米；2004～2012年，出省年平均径流量207.6亿立方米，年平均增加5.7亿立方米；2015年出省径流量为158.01亿立方米。2004～2012年，澜沧江出省年平均径流量107亿立方米；2015年出省径流量为102.51亿立方米。到2015年，三江源地区地表水水质总体状况为优（见图3.2）。

### （三）水源涵养功能显著提升

1997～2004年三江源地区林草生态系统多年年平均水源涵养量

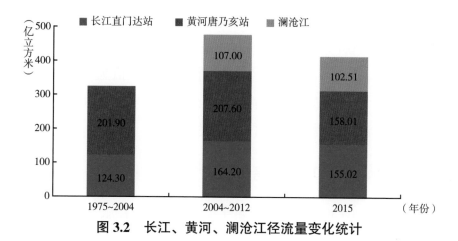

图 3.2 长江、黄河、澜沧江径流量变化统计

为 142.49 亿立方米，2004～2012 年林草生态系统多年年平均水源涵养量为 164.71 亿立方米，增加了 22.22 亿立方米。面积大于等于 1 平方公里的水体 226 个，总面积为 5785.5 平方公里，比 2006 年增加了 261.25 平方公里，其中扎陵湖和鄂陵湖面积分别增加 32.69 平方公里和 64.36 平方公里，增幅分别为 6.47% 和 11.03%。2015 年，三江源地区草地土壤水分含量均值在 7.5%～15.5%。

（四）生态系统结构逐渐向良性方向转化

主要表现为湿地生态系统面积扩张，荒漠生态系统逐步得到保护和恢复，草地退化态势得到遏制。2004～2012 年，三江源地区内森林面积增加 15.3 平方公里，草地面积增加 123.7 平方公里，水体和湿地面积增加 279.85 平方公里，荒漠面积减少 492.61 平方公里。

（五）区域气候发生变化

1975～2004 年三江源地区各气象站点年平均气温为 -0.58℃，

年平均气温变化率约为0.38℃/10年；2004~2012年各气象站点年平均气温为0.4℃，年平均气温变化率约为0.1℃/10年，增温速率明显降低；1975~2004年各站点年降水量均值为470.62毫米，2004~2012年各站点年降水量均值为518.66毫米，年均降水量增加48.04毫米，湿润指数平均增加5.03左右。

### （六）水土保持功能提高

1997~2004年三江源地区多年年平均土壤保持服务量为5.46亿吨，2004~2012年年平均土壤保持服务量为7.23亿吨，增加了1.77亿吨。

### （七）生物多样性增加

自然保护区内藏野驴、岩羊、野牦牛等野生动物种群数量明显增多，栖息范围呈扩大趋势。高原植物种群和土著鱼类等水生生物的多样性得到有效保护，生物多样性逐步恢复。

### （八）天然草地放牧压力减轻

截至2013年，自然保护区内共实现天然草地减畜342万羊单位，牲畜超载率降低了41.94个百分点，天然草地放牧压力明显减轻。另外，通过生态移民从自然保护区核心区转移1.07万户5.39万人，降低了牧民对天然草地的利用强度。

### （九）生态防护体系和监测网络基本健全

通过建立和加强林地草地生态系统、自然保护区保障体系及生态和环境监测网络，项目区的生态防护和监测能力明显增强。

## 三　三江源生态保护综合试验区二期规划

青海省委、省政府坚决贯彻落实党中央、国务院决定，始终将生态保护作为重要责任，将《青海三江源生态保护和建设二期工程规划》编制作为头等大事来抓。二期规划后经青海省人民政府第100次常务会议和十二届省委第四次常委会审议，于2012年10月上报国家发改委。国家发展和改革委员会征求了科技部、财政部、国土资源部、环境保护部、水利部、农业部、国家林业局、中国气象局等部门意见，并委托中国国际工程咨询公司进行了评估。根据评估报告和各部门意见，进行了修改完善。国家发改委于2014年1月8日批准了《青海三江源生态保护和建设二期工程规划》，《规划》总投资160.57亿元，建设内容共两大类24项工程，同时正式启动实施二期规划。

二期规划是一期工程的延续、拓展和提升，是落实《总体方案》的主要载体，是统筹生态保护、民生改善、区域发展的主要支撑。尽管一期工程实施以来，三江源国家级自然保护区内生态状况恶化趋势有所好转，植被覆盖度有所提高，生物多样性有所改善，但三江源区生态系统功能恢复是一个长期过程，群落结构、物种组成和土壤养分仍处于进化与退化的竞争阶段。需要进一步实施生物与工程措施，巩固已取得的生态建设成果。因此，实施二期规划，有利于进一步采取更加有力的措施解决好三江源生态环境问题。

### （一）二期规划范围

二期规划范围为青海三江源国家生态保护综合试验区，包括玉树藏族自治州、果洛藏族自治州、海南藏族自治州、黄南藏族自治

州全部行政区域的21个县和格尔木市的唐古拉山镇，共158个乡镇、1214个行政村（含社区），总面积为39.5万平方公里。

### （二）二期规划目标

二期规划基准年为2011年，规划期限为2013~2020年。

二期规划主要目标：到2020年，林草植被得到有效保护，森林覆盖率由4.8%提高到5.5%，草地植被覆盖度平均提高25~30个百分点；土地沙化趋势得到有效遏制，沙化土地治理率达到50%，沙化土地治理区内植被覆盖度达30%~50%；水土保持能力、水源涵养能力和江河径流量稳定性增强；湿地生态系统状况和野生动植物栖息地环境明显改善，生物多样性显著恢复；农牧民生产生活水平稳步提高，生态补偿机制进一步完善，生态系统步入良性循环。

### （三）二期规划工程功能分区

《总体方案》根据综合试验区的自然条件、资源环境承载能力、经济社会发展情况和区域功能定位，将三江源试验区划分为重点保护区、一般保护区和承接转移发展区，本规划分区布局与试验区保持一致。依据各分区特性，优化空间布局，实施分类指导，确定重点保护区域和主要任务，以努力实现总体目标（见表3.2）。

**1. 重点保护区**

重点保护区是指在构成三江源地区生态安全格局中发挥特殊重要作用，以生态环境保护为核心、原则上禁止从事开发经营活动的区域。

（1）区域范围。以国家级自然保护区为主，包括三江源、可可西里、隆宝3个国家级自然保护区，以及年宝玉则、坎布拉、贵德等国家地质公园、森林公园、湿地公园及风景名胜区，面积19.8万平方公里，占规划区总面积的50.1%。

（2）区域功能。该区域高寒草原草甸、冰川、沼泽湿地及河流、湖泊广布，寒温带针叶林错落分布，野生动植物多样性丰富，主要承担水源涵养及产水功能，是发挥生态功能的核心区域。该区域生态环境脆弱，一旦遭到破坏，恢复难度极大，必须依据国家有关法律法规，采取严格的生态环境保护措施加强保护和修复。

2. 一般保护区

一般保护区是指在构成三江源区域生态安全格局中发挥维护生态系统完整性基础作用，优先保护生态环境，并根据资源环境承载能力适当发展畜牧业生产经营活动的区域。

（1）区域范围。包括重点保护区和承接转移发展区之外的地区，总面积18.91万平方公里，占规划区总面积的47.9%。

（2）区域功能。该区域是高寒草甸草原、干旱半干旱草原的主要分布地区，生物多样性较为丰富，也是从事草地畜牧业等生产经营活动的传统地区，具有一定的资源环境承载能力。加大保护建设力度，控制草原利用强度，实现草畜平衡；合理布局牧区聚落，引导牧民有序集聚；促进生态系统修复和恢复，努力实现生态平衡和人与自然和谐。

3. 承接转移发展区

承接转移发展区是指在生态保护建设和城镇化进程中，具备一定承接转移农牧业人口和产业发展潜力，需要统筹开发开放与生态保护的区域。

（1）区域范围。包括共和、贵德、尖扎、同仁等县境内黄河谷地地区，以及规划区州府、县城所在地和重点小城镇，基本呈点状分布，总面积0.8万平方公里，占规划区总面积的2%。

（2）区域功能。该区域生态类型复杂多样，农牧交错，耕地相对集中，灌区开发条件较好，资源环境承载能力相对较高，是发展城镇、集聚产业和人口的主要地区。要在切实做好生态保护建设前提下，构建以州府县城为中心、重点乡镇为基础的城镇体系，承接重点保护区和一般保护区的转移人口，改善人居环境，提升社会保障和服务功能；转变发展方式，科学规划黄河沿岸综合开发，加强农牧业基础设施建设；因地制宜地推动产业向园区集中，有序发展特色优势产业，努力实现低碳绿色发展。

表 3.2　二期规划工程功能分区

| 功能分区 | 含义 | 区域范围 | 区域功能 |
| --- | --- | --- | --- |
| 重点保护区 | 在构成三江源地区生态安全格局中发挥特殊重要作用，以生态环境保护为核心、原则上禁止从事开发经营活动的区域 | 以国家级自然保护区为主，包括三江源、可可西里、隆宝3个国家级自然保护区，以及年宝玉则、坎布拉、贵德等国家地质公园、森林公园、湿地公园及风景名胜区，面积19.8万平方公里，占规划区总面积的50.1% | 高寒草原草甸、冰川、沼泽湿地及河流、湖泊广布，寒温带针叶林错落分布，野生动植物多样性丰富，主要承担水源涵养及产水功能，是发挥生态功能的核心区域。该区域生态环境脆弱，一旦遭到破坏，恢复难度极大，必须依据国家有关法律法规，采取严格的生态环境保护措施加强保护和修复 |
| 一般保护区 | 在构成三江源区域生态安全格局中发挥维护生态系统完整性的基础作用，要优先保护生态环境，并根据资源环境承载能力适当发展畜牧业生产经营活动的区域 | 包括重点保护区和承接转移发展区之外的地区，总面积18.91万平方公里，占规划区总面积的47.9% | 是高寒草甸草原、干旱半干旱草原的主要分布地区，生物多样性较为丰富，也是从事草地畜牧业等生产经营活动的传统地区，具有一定的资源环境承载能力。加大保护建设力度，控制草原利用强度，实现草畜平衡；合理布局牧区聚落，引导牧民有序集聚；促进生态系统修复和恢复，努力实现生态平衡和人与自然和谐 |

续表

| 功能分区 | 含义 | 区域范围 | 区域功能 |
|---|---|---|---|
| 承接转移发展区 | 在生态保护建设和城镇化进程中,具备一定承接转移农牧业人口和产业发展潜力,需要统筹开发开放与生态保护的区域 | 包括共和、贵德、尖扎、同仁等县境内黄河谷地地区,以及规划区州府、县城所在地和重点小城镇,基本呈点状分布,总面积0.8万平方公里,占规划区总面积的2% | 生态类型复杂多样,农牧交错,耕地相对集中,灌区开发条件较好,资源环境承载能力相对较高,是发展城镇、集聚产业和人口的主要地区。要在切实做好生态保护建设前提下,构建以州府县城为中心、重点乡镇为基础的城镇体系,承接重点保护区和一般保护区的转移人口,改善人居环境,提升社会保障和服务功能;转变发展方式,科学规划黄河沿岸综合开发,加强农牧业基础设施建设;因地制宜地推动产业向园区集中,有序发展特色优势产业,努力实现低碳绿色发展 |

## (四)二期规划建设内容

生态保护和建设工程主要内容如表 3.3 所示。

表 3.3 生态保护和建设工程主要内容

| 序号 | 类别 | 建设内容 |
|---|---|---|
| 一 | 草原生态系统保护和建设 | 禁牧封育和草畜平衡管理、退牧还草、黑土滩治理、草原有害生物防控 |
| 二 | 森林生态系统保护和建设 | 现有林管护、封山育林、人工造林、农田防护林更新改造、中幼林抚育、林木种苗基地建设、森林有害生物防控 |
| 三 | 荒漠生态系统保护和建设 | — |
| 四 | 湿地、冰川与河湖生态系统保护和建设 | 水土保持、湿地和雪山冰川保护、人工影响天气、饮用水源地保护 |
| 五 | 生物多样性保护和建设 | 湖泊湿地禁渔、鱼类增殖放流、濒危野生动物监测 |
| 六 | 支撑配套工程 | 生态畜牧业、农村能源建设、生态监测、基础地理信息系统、科研和推广、培训、宣传教育 |

## 四 社会各界在行动

三江源生态保护事关中国、亚洲甚至北半球的生态安全和中华民族的长远利益，在当前全球大气变暖、生态急剧退化的总趋势下，三江源生态环境面临严峻形势，引起社会各界高度关注。三江源的生态保护不仅是国家和政府的事，而且与每一位中华儿女都息息相关，需要全社会的共同努力。

近年来，随着社会经济的发展，人们的环境保护意识在不断增强，各界人士、各类团体以及国际组织参与三江源生态保护的积极性也在不断提高。例如，2001年3月12日爱立信（中国）有限公司为保护可可西里藏羚羊捐赠了交通、通信、物资设备和200万元的现金支票，由于三江源重要的生态地位，这类捐赠将越来越多。为了给国内外关心支持三江源生态保护事业的公益人士提供一个开放可靠的平台，三江源生态保护基金会于2012年10月22日正式成立。基金会严格遵守国家相关法律法规以及基金会法定规范，建立科学的内部管理结构和议事规则，并按基金会章程办事，确保规范运行，管好用好每一笔资金，成为三江源国家生态保护综合试验区的有力推动者。基金会成立以来，多次开展促进三江源生态保护的募捐活动，接受海内外热心环保公益事业的有关组织和个人的捐助，为捐赠方无偿提供完善、规范的相关服务，有效地落实各项捐助意愿和捐赠资金，切实把社会各界对三江源生态保护事业的一片爱心落到实处。所受捐款主要用于资助三江源资源与生态保护项目和科学研究、科技开发，开展和促进三江源生态保护相关的宣传教育、学术交流以及国际交流与合作等活动，使社会公益基金能够充分投入到三江源生态保护事业中。例如，基金会资助的"公益宣传平台建设""长江黄河土著鱼类增殖放流""草原综合治理示范区"等重要项目，在三江源生态保护的宣传教育和信息公开、生物多样性维持和濒危动植物的生存保障、草原生态综合治理等方面都已取得很好的成果。

除三江源生态保护基金会，一些完全由公众自发组织的公益性机构，如三江源生态保护协会，也为三江源的生态保护做出了重要贡献。三江源生态保护协会的前身是玉树州三江源生态环境保护协会，2008年4月9日重新在青海省民政厅登记注册并改为现在的名称。三江源生态保护协会以藏族成员为主，致力于青藏高原地区生态环境与传统优秀生态文化的保护与宣传，关注青藏高原地区的可持续发展。

三江源是一块科研的宝地，由于其所处位置极为特殊，形成了独具特色的自然风貌和人文风俗，具有丰富的科研价值。同时，三江源的生态保护工作面临许多科学性难题亟待解决。近年来，众多高等院校和科研单位投入三江源的科学研究工作中，有关三江源的学术论文和专著数量不断增加，科研成果显著。2010年10月28日青海大学－清华大学三江源研究院成立。该研究院由青海大学和清华大学两校共同发起组成，重点组织两校有关科研团队，联合国内外高校和研究机构的科技资源共同开展三江源科研合作，在三江源生态保护、后续产业发展、公共服务体系建设可持续发展、新能源综合利用及野外台站建设等相关领域开展全面、系统、有针对性的研究。同时，召开学术论坛，举办各类研讨会，出版研究报告，以此推动三江源生态恢复、保护和建设工作，为政府提供咨询服务，力争成为国内知名、国际有一定影响力的研究机构（见图3.3）。

中国科学院地理与资源研究所成立了三江源生态研究课题组，经过8年的研究，课题组借鉴联合国新千年生态系统评估（MA）的理

**图3.3 青海大学-清华大学三江源研究院**

论框架,以生态环境的有效连续监测评估和生态建设成效评估为核心目标,以空间信息技术为主要手段,建设了三江源生态环境综合数据库系统,设计构建了生态系统综合评估指标体系,研发了地面观测-生态模型-遥感对地观测的天地一体化生态监测技术体系,提炼了20世纪70年代以来三江源生态系统格局和服务功能的变化规律,为生态工程成效评估建立了科学完善的三江源区域生态环境的动态过程本底,并于2012年11月出版了《三江源生态系统综合监测与评估》一书,总结和归纳了重要研究成果。三江源的科学研究为该地区的生态保护和恢复提供了科学依据。

2016年6月20~28日,青海省水利厅和青海大学联合组织清华大学、青海省水文水资源勘测局、河南黄河水文勘测设计院、北京东方园林股份有限公司、德清北辰信息技术有限公司、西安山脉科技发展有限公司、北京美科华仪科技有限公司等单位的27名专业技术人员,组成三江源国家公园综合科学考察队(见图3.4)。

**图 3.4　三江源国家公园综合科考队**

科考队旨在加强源区水文监测工作，探索地区水文变化规律，提高源区水文水资源支撑服务能力，更好地为"中国三江源国家公园"建设及"青海可可西里申遗"工作提供技术支撑。科考队从西宁出发，经过同仁、玛多、玉树、囊谦、杂多、治多、曲麻莱等三江源头地区，对长江南源当曲、北源楚玛尔河、正源沱沱河及干流通天河和澜沧江源区扎阿曲、扎那曲、子曲等河流，以及现有下拉秀、香达、雁石坪水文巡测站开展实地考察，以期获得长江源区和澜沧江源区的河流水系、气象、水文、河道河势、水资源、水生态环境、地形地貌、源区社会经济、通信条件等基础资料；通过对源区的实地勘查，课题组可获得原始信息，以便制订新建水文测站选址和改建站升级改造方案，进而完善和优化源区水文监测站网；调查了解源区水文生态状况；基于水汽通量研究现场考察适宜的空中调水作业区。此次考察将有利于课题组进一步掌握长江源头和澜沧江源头的生态环境现状，为大江大河上游的开发与保护提供科学指导。

为了深入了解三江源生态环境状况,在2012年以来历年科学考察的基础上,2017年5月31日至6月7日,长江科学院、青海省水利厅和青海大学联合多学科综合科考队共34人,完成以各拉丹东雪山为核心,全面覆盖长江正源沱沱河、南源当曲、北源楚玛尔河和澜沧江源头的水文水资源、水生态环境、河流水系、水生生物和底栖生物、冻土冰川和地形地貌等多学科综合科学考察任务(见图3.5)。

图3.5　囊极巴陇科考队纪念碑合影

# G.4
# 三江源区生态环境保护工作展望

习近平总书记指出，河川之危、水源之危是生存环境之危、民族存续之危。必须从全面建设小康社会、实现中华民族永续发展的战略高度来重视解决水安全问题。国务院总理李克强在第十二届全国人民代表大会第三次会议的政府工作报告中指出，森林草原、江河湿地是大自然赐予人类的绿色财富，必须倍加珍惜。要推进重大生态工程建设，拓展重点生态功能区，办好生态文明先行示范区，开展国土江河综合整治试点，扩大流域上下游横向补偿机制试点，保护好三江源。国家对生态文明建设的新部署、新要求为进一步加强三江源生态保护指明了前进方向，提供了难得机遇。

三江源最大的价值在生态、最大的潜力在生态、最大的责任也在生态。三江源生态保护和建设再起航，站在新的历史起点上，三江源区的生态环境保护工作任重道远。

## 一 三江源保护面临的挑战与问题

三江源生态环境保护与建设取得阶段性成果，成效明显，但仍存在一些问题，三江源生态环境保护和建设任重道远。

一是生态整体退化趋势未得到根本遏制。通过前期的保护和治理，生态退化趋势得以缓解，但是仍有80%的黑土滩退化草地和

60%的沙化土地未治理，38%的天然草地未进行退牧还草，鼠害在已治理的草地上也有复发的现象。整个三江源区中度以上退化草地占可利用草地面积的50.4%，其中，黑土滩型退化草地约占39%；土地沙化趋势依然存在，沙化土地面积约占规划区的8%；水土流失面积占规划区的30.7%；鼠虫害发生面积仍有16.7万平方公里，占规划区的42.3%。大面积的退化草地、沙化土地和鼠虫害草地，对三江源区生态系统造成严重威胁，生态整体退化趋势尚未得到根本遏制。

二是超载过牧现象仍然存在。导致三江源区生态退化的最主要的因素是超载过牧。前期针对超载问题，政府安排了退牧还草、建设养畜及生态移民工程，但由于对转变农牧民生产方式引导不足，舍饲畜牧业基础设施建设薄弱，牧业生产仍以传统放牧为主，减畜手段以强制性为主，退牧还草饲料补助、草原奖补机制等补助也只能满足农牧民基本生活水平，难以保障实现脱贫致富的目标，至今超载问题尚未得到根本解决。三江源区现有可利用天然草地26.37万平方公里，其中中度以上退化草地13.30万平方公里，未退化和轻度退化草地13.07万平方公里，实际载畜量1828.84万羊单位。若中度以上退化草地全面实施禁牧，未退化和轻度退化草地、现有人工草地、改良草地的理论载畜量只有926.63万羊单位，草畜矛盾突出，禁牧减畜任务繁重。

三是自然条件和技术制约依然严重。三江源区平均海拔约4000米，气候寒冷、空气稀薄、紫外线强、地形复杂、降水时空分布不均，冻土广布，土壤发育过程和植被生长缓慢，破坏后极难恢复，特殊的气候地理条件和敏感脆弱的生态环境特征决定了生态恢复和治理的技术难度远远高于低海拔地区。目前，针对高寒地区特殊的自然生态系统的保护和恢复技术仍存在不成熟甚至空白的领域，制约了生态保护和建设工程效益的发挥。比如黑土滩治理、湿地保护、野生动物保护和沙化土地治理等技术方法比较单一。尽管在一期列入一批科研项目，国家也安排了相应科技支撑项目，但部分科研成果应用效果不明显。例如，对不同立地条件下的黑土滩治理和人工草地建植，以及不同地区各类鼠害防治等技术的研究不够全面和成熟，难以满足工程建设需求。

四是生态保护和民生改善矛盾突出。随着生态保护和建设的继续实施，综合试验区建立后范围扩大，区域内矿产资源、水电资源、中藏药资源的开发利用将受到进一步限制，草原禁牧和限牧面积进一步增加，势必影响当地经济的发展和农牧民生活水平的提高。而与此相对应的是人口增加，预计到2020年人口将达到134万；基础设施依然薄弱，公共服务能力低于全省平均水平；产业结构调整难度大，生态经济短期难以形成规模，牧民转产转业制约多，人口与环境承载能力之间的矛盾日益突出。加之交通不便、运输困难、施工期短等因素，更增加了生态建设工程的实施难度和工程建设成本。如何积极探索形成有利于生态保护、民生改善、经济发展、社会进步相协调的生态保护管理体制和规范长效的生态补偿机制，是今后面临的重要任务和挑战。

## 二 实施好三江源生态保护和建设二期工程规划

《青海三江源生态保护和建设二期工程规划》是一期工程的延续、拓展和提升，是统筹生态保护、民生改善、区域发展的主要支撑。总结一期规划实施的经验，在实施二期工程规划时要重点做好以下几方面工作。

一是要坚持规划引领，始终用生态文明建设的总体要求来统领整

个生态保护工作。三江源生态系统的稳定事关长江、黄河、澜沧江中下游地区经济社会可持续发展。实践证明，总体规划对促进三江源区生态保护和科学发展发挥了重要作用。青海省委、省政府按照生态文明建设的总要求，确立了生态立省的战略，树立了筑牢国家生态安全屏障的大局意识，坚持总体规划确定的指导思想和基本原则，加强领导，统一协调，分级负责，精心组织实施规划项目，为实现三江源区生态系统改善打下了坚实基础。

二是加强组织领导，进一步强化责任落实。为实施好规划项目，青海省成立了省、州、县三级领导小组和办事机构，按工程项目性质分别成立了农牧、林业、生态移民、技术咨询、生态监测等8个工作组，项目区实行县、乡、村、户分级承包责任制，各州、县主要领导为第一责任人，有力地保障了工程的顺利实施。下一步，各级党委、政府要把三江源生态保护作为实施生态立省战略和转变发展方式的切入点，进一步加强组织领导，全面落实工作责任，尤其要将生态保护和建设列为三江源区各级政府工作的主要考核内容，形成一级抓一级、齐抓共管、相互配合、合力推进的良好氛围。

三是积极探索，加强制度创新。在工程管理、项目建设、生态社区管理、扶持生态移民后续产业、健全草畜平衡保障机制等方面，要严格按照有关政策规定执行。同时，要积极探索，加强制度创新，为规划项目的实施提供有力支撑。比如青海省在一期规划实施过程中，在全国率先制定了《关于探索建立三江源生态补偿机制的若干意见》，先行先试，确定了11项补偿政策，极大地调动了项目区群众参与生态保护的积极性；设立了三江源生态保护后续产业发展基金，成立了三江源生态保护基金会，充分利用市场资源，多渠道筹资和集约利用社会各方资金用于三江源区的资源补偿、生态补偿，统筹解决生态环境保护建设、区域经济社会发展、保障和改善民生、实现公共服务均等化等突出问题；推行绿色绩效考核等。

四是处理好保护与发展的关系，实现科学保护与绿色发展相统一。在三江源生态保护工作中，要始终坚持自然修复为主、人工治理为辅的原则，从超载过牧、开垦草原、滥采乱挖、偷猎盗伐等人类不合理活动入手强化管理，同时通过鼠害防治、草地围栏、人工草地建设和天然草地改良、封山育林、湿地保护等措施促进生态系统逐步改善。要始终坚持"保护中发展，发展中保护"的原则，先易后难，稳步推进，要落实好各项惠民利民政策，结合生态保护工程加强基础设施建设，加强生态移民社区供排水、供电、道路、教育、卫生等基础设施建设，大力扶持生态移民后续产业，确保生态移民搬得出、稳得住、能致富，增强三江源区生态保护的内生动力，实现生态保护和经济社会协调发展。

## 三 进一步加大科技投入和科学研究力度

提高三江源区生态保护的科学技术支撑能力是搞好三江源区生态环境保护的重要环节。

在一期规划实施过程中，重点加强了以生态保护为重点的应用性科研项目研究和推广。先后开展生态保护科学研究及应用推广项目96项，其中"三江源湿地变化与修复技术研究"和"三江源退化草地生态系统恢复治理研究"等成果达到国际领先水平。制定了黑土滩退化草地分级标准，查清了三江源黑土滩退化草地的面积、类型及分布，提出了黑土滩治理方案，建成了退化草地治理信息系统，初步确定了不同类型退化草地形成原因和恢复机理，提出了可持续利用的对策和模式。实现了中华羊茅、青海冷地早熟禾等优质牧草本地化扩繁扩育，为退化草地的治理提供了重要支撑。强化监测站点和监测队伍建设，三江源生态监测和评估体系不断完善，成为青藏高原生态监测的典型示范。结合工程实际需要，培训管理干部6000多人次、农牧民近5万人次，确定科技示范户1700多户，这些措施不仅为工程措施顺利实施提供了有力的科技支撑，更重要的是培养了一大批管理人才和转产致富能手，从而增强了三江源区科技成果的整体推广效果，提高了保护水平。

为进一步加强对三江源区生态保护的科学研究和关键技术攻关，下一步的工作主要包括以下方面。

一是加强生态监测。完善生态环境监测、信息传输、预警服务、监测基地技术保障、评价服务五大系统，构建生态环境遥感解析中心、生态环境地面观测站等生态环境监测工作平台，建立生态环境数据库和信息查询系统，完善信息资源共享和会商分析机制，实行年度评估综合报告制度，提高监测、预警和评估能力，定期对生态环境状况、生态系统结构、生态功能、生态敏感性、资源环境承载能力以及生态恢复效果等进行全面评估。

二是深入开展基础科学研究。包括三江源区生态系统演替机理、气候变化与生态系统保护的关系、三江源区重要生态指标定量分析等，着力提高对三江源区生态环境变化的科学认知和判断，找准变化机理，提出切实可行的治理措施。

三是加大应用技术研究和推广力度。开展三江源区草地、湿地、林地、荒漠化土地修复治理，生态畜牧业、有机畜牧业等关键技术攻关研究，积极开展生态保护、环境治理、产业发展等先进技术试验示范，加强对牧民和技术人员的技能培训，让技术项目逐步推广到整个三江源区，提升科技水平。

国务院各有关部门、三江源区各级政府，应该通过各种有力措施，增加人才、资金等方面的投入，积极支持三江源区的科学技术研究和应用，为三江源区的生态环境保护和建设提供有力的支撑和保障。

## 四 建立全社会参与三江源区保护的机制

三江源区是少数民族聚集区，民族构成以藏族为主，占90%左右，宗教氛围浓厚，完全依赖当地草原生态环境的畜牧业是该区域的主导产业。作为拥有游牧文化传统的藏民族，自古以来其生态文化和利用自然资源的乡土知识就在传统的生态保护中发挥着积极的作用。当地政府应调动当地群众参与的积极性，增强公众参与环境保护的意识，制定公众成功参与的激励模式。

公众参与，一方面能够促进政府和群众在三江源区生态环境保护和建设方面达成共识，凝聚保护力量；另一方面在生态环境保护方面对群众可以起到良好的教育作用，提高全社会的环境意识，从而使群众对三江源区生态环境的保护变成一种自觉行为。

由于三江源区生态环境保护的着眼点不限于区域本身，更重要的是考虑到黄河、长江和澜沧江中下游乃至全国的水安全和生态安全。

为引导社会各界及当地民众积极参与源区生态保护，主要应做好以下几个方面的工作。

首先，在群众参与主体上，应强化民间环境组织的作用。民间环境组织不同于一般的单位和个人，其在生态环境保护方面具有更大的作为。比如2012年成立的三江源生态保护基金会是一个纯公益性的公募基金会，通过开展促进三江源生态保护的募捐活动、资助三江源资源与生态保护项目、开展国际交流合作等多种形式对三江源区的生态环境保护起到了巨大的推动和保障作用。

其次，在公众参与范围上，应进一步放宽限制，实行全过程参与，既要有事前决策过程的参与，也要有执行过程的参与，还包括事后监督过程的参与。决策过程的群众参与主要解决的是三江源区生态环境保护和建设中涉及当地群众重大利益的事项，在其政策和法律文件制定过程中要注意听取当地群众的意见，以保证其顺利实施。执行过程的群众参与主要解决的是政府监管能力不足的问题，目的是通过群众参与，配合政府监管，提高效率。事后监督过程的群众参与主要是落实责任追究制，实现三江源区生态环境保护和建设目标。

最后，要进一步加强宣传教育。要通过广播、电视、电影、报纸、互联网等多种渠道加强三江源生态保护和建设的宣传，通过全方位、多角度的报道宣传，进一步提升社会各界对三江源生态保护的关注度，进一步提高对三江源生态保护重要性的认识，着力营造各界支持、热情帮助、广泛参与的良好氛围。

# 参考文献

秦大河:《三江源区生态保护与可持续发展》,科学出版社,2014。

中国科学院地理与资源研究所:《三江源生态系统综合监测与评估》,科学出版社,2012。

## 鸣　谢

青海省水利厅
青海省气象局
青海省林业厅
青海省环保厅
三江源国家公园管理局
青海省三江源生态保护和建设办公室等部门
为本卷提供了相关数据资料，并审核把关，在此一并感谢！

# The Editorial Committee of *A Research Report on Sanjiangyuan Ecological Protection (2017)*

**Dean:** Guang-qian Wang

**Counselor:** Guang-qian Wang, Hong-zhong Yao

**Chief Editor:** Jia-hua Wei, Tie-jian Li

**Committee Members (in Pinyin)**

Jun Cao, Gang Chen, Wang Fu, Rui-jun Huang, Yue-fei Huang,

Feng-xia Li, Tao Liu, Xi-ning Liu, Zhi-qiang Ma, Hai-hong Su,

Zhi-min Wang, Zhong-jing Wang, He-quan Zhang, Gui-yun Zheng

**Written by:** Wang Fu

**Translated by:** Rong-ling Fan, Qingyu -Han, Ping Liu, Qin Zhang

# About the Authors

**Jia-hua Wei**, Ph. D., is the Changjiang Scholar Distinguished Professor of the Ministry of Education. Since 2004, he has been working in the Department of Hydraulic Engineering, Tsinghua University. During December 2007 to January 2009, he worked as a visiting scholar at Cornell University in the United States. Since 2014, he has been working for the supporting program of Tsinghua University to Qinghai University, as a deputy dean of the School of Hydraulic and Electric Power of Qinghai University and a chief scientist of hydrology and water resources at the State Key Laboratory of the Sanjiangyuan Ecology and Plateau Agriculture.

Professor Wei's research interests are in the areas of hydrology and water resources, hydroinformatics, surface water-groundwater interactions, atmospheric water resources utilization, etc. He has led/participated in more than 40 research projects from the National Natural Science Foundation, the National Key Technology R&D Program, the National Key Research Program, the Special Program for Scientific Research of Public Welfare, Major Projects Consultation, and others. His publications include more than 120 journal articles and conference papers, and 5 edited monographs. His honors and awards include the $10^{th}$ "China Youth Award of Science and Technology" which was jointly sponsored by the Organization Department of the Communist Party of China, Ministry of Personnel, and China Association for Science and Technology. He won the $2^{nd}$ Place of National Science and Technology Progress Award. He also won the $1^{st}$ Place for five times and the $2^{nd}$ Place for 3 times of the Provincial Scientific and Technological Progress Award.

**Tie-jian Li**, Ph.D., is an Associate Professor of River Research Institute of the Water Conservancy Department, Tsinghua University, a Young Scholar of Changjiang Scholars Programme, and a Visiting Professor of Kunlun Scholar in the Department of Water Conservancy and Electric Power in Qinghai University. He obtained his bachelor's and doctoral degrees in 2003 and 2008, respectively, in the Water Conservancy Department, Tsinghua University. Since 2010, he has been working in Tsinghua University, and from 2014 he has been working for the supporting program of Tsinghua University to Qinghai University.

Professor Li mainly focuses on the study of River Dynamics and Hydroinformatics. His research interests include the Dynamic Model of Soil Erosion and Sediment Transport in River Basins, River Basin Simulation Technology, River Basin Geometry, Atmospheric Water Vapor Channel, Cloud Water Resources Analysis, Big Data Application to River Systems, and others. He established the Yellow River Digital Model and the Digital Platform of River Basin, extracted the High Resolution Global Digital River Network Hydro30, and carried out pioneering research on the Big Data Application to River System. His publications include more than 50 papers, among which more than 20 are listed in SCI, and there are more than 200 times of citations of his papers in SCI. His honors and awards include the $2^{nd}$ Place of Science and Technology Progress Award of the Ministry of Education ($2^{nd}$ completed person) in 2015. His monograph entitled *Dynamic Model of Soil Erosion and Sediment Transport in River Basins* was awarded the Second China National Book Award in 2010. He is a member of many academic organizations, such as IEEE, and also a reviewer for many highly reputed international peer-reviewed journals, including *Water Resources Research*, *Journal of Hydrology*, and *Hydrological Processes*.

# Abstract

Sanjiangyuan is the headwater region of the Yangtze River, the Yellow River and the Lancang River that flows through six countries. It is located in the south of Qinghai Province, the northeast of the Qinghai-Tibetan Plateau, and the area lies between 31°39'-36°12' North, 89°45'-102°23' East with an average height of 3500-4800 meters above sea level. According to the administrative region delimited by the Sanjiangyuan National Ecological Comprehensive Protection Pilot Zone, the Sanjiangyuan region covers 21 thorough administrative counties including the four Tibetan Autonomous Prefectures of the Yushu Prefecture, the Guoluo Prefecture, the Hainan Prefecture and the Huangnan Prefecture; as well as the Tangula Township of the Geermu City in the Haixi Mongolian and Tibetan Autonomous Prefecture with a total area of 395,000 km2. According to the river basin boundary, the basin area of the Yellow River is 122,000 km2 (far western part of the Tangnaihai Hydrologic Station) which accounts for 39% of the total area of the Sanjiangyuan Region. The basin area of the Yangtze River is 137,700 km2 (far western part of the Zhimenda Hydrologic Station), which accounts for 44.40% of the total area of the Sanjiangyuan Region. The basin area of the Lancang River is 52,900 km2 (far western part of the Changdu Hydrologic Station), which accounts for 16.90% of the total area of the Sanjiangyuan Region.

Sanjiangyuan is the source area of rivers and water resources, the ecological protection of Sanjiangyuan is undoubtfully beneficial from the unique hydrological and water resources condition in this area. Therefore, hydrological conditions and water resources is listed as the first volume in the Green Paper's publication plan. The book is divided into general report and subject report. The general report briefly introduces the natural and socio-economic characteristics of the Sanjiangyuan Region. The subject report introduces the hydrological conditions and water resources; the course and achievements of Sanjiangyuan Ecological Protection; the outlook of Sanjiangyuan Ecological Protection separately. Hydrological condition and water resources is the core part of the book. Based on the existing data of Qinghai Province, this chapter talks about new information achieved through the technology of satellite precipitation inversion and the Tsinghua Hydro30, a HR worldwide river network that is established by Tsinghua University independently as complements to have a holistic and more realistic understanding of the hydrological and water resources conditions in the Sanjiangyuan Region.

Chapter One "Natural and Socio-economic Characteristics of the Sanjiangyuan Region" firstly introduces the geographical location of the Sanjiangyuan Region, clarifying the concept and origins of the source area of the Sanjiangyuan Basin, the Sanjiangyuan National Nature Reserve, the Sanjiangyuan National Ecological Protection Pilot Zone and the Sanjiangyuan National Park. The book takes the Sanjiangyuan National Ecological Protection Pilot Zone as the research object, since the pilot zone is the largest in area which is more convenient for research. In addition, the source area of the Sanjiangyuan Basin is taken as a reference. Following topography and terrain features, climate, surface cover features, species and biological communities are introduced briefly. Finally by this chapter, social and economic development is introduced based on population and industry development.

Chapter Two "Hydrological Conditions and Water Resources" introduces precipitation characteristics, river systems, wetland resources, glacier snow mountains and groundwater resources one by one. Due to the scarcity of surface gauging stations in this region, the book takes advantages of satellite remote sensing rainfall data spatial coverage, analyzes the temporal and spatial distribution characteristics of precipitation by using surface gauging station data and satellite remote sensing data, and conducts satellite data contrast verification. The analysis of river system mainly adopts the data of the

# Abstract

Tsinghua Hydro30 digital river network, and explains the general situation of the water system in the Yellow River Source Area, the water system in the Yangtze River Source Area, the water system in the Lancang River Source Area, inflow rivers, lakes, and main tributaries respectively. This part also introduces some rivers cross-sectional conditions and historically measured runoff data at major hydrological stations. At the same time, this chapter gives a brief introduction of wetland resources, glacier snow mountains and groundwater resources.

Chapter Three "The Course and Achievements of the Sanjiangyuan Ecological Environment Protection" introduces the establishment process of the Sanjiangyuan Nature Reserve, the Sanjiangyuan Comprehensive Pilot Zone and the National Park. The obvious implementation effect of the General Plan is discussed in vegetation coverage, runoff volume of rivers and biodiversity and so on. At the same time, this chapter also brings brief introductions of the Phase II Plan of the Sanjiangyuan Comprehensive Ecological Protection Pilot Zone and the actions of community in the society to the Sanjiangyuan Region protection.

The ecological protection and construction of the Sanjiangyuan have achieved remarkable results; the ecological system structure has been partially improved; the deterioration of the grassland has been basically curbed; the contradiction between the grassland and grazing has been lessened; the ecological function of the wetland has improved step by step; the lakes have been enlarged and the water provision ability has been strengthened. However, the ecology protection of this region still has its challenges and problems. The last chapter of this book "The Outlook of the Sanjiangyuan Ecological and Environment Protection" first briefly describes the challenges and problems Protecting Sanjiangyuan. In order to propose prospects and suggestions for the protection of the Sanjiangyuan region, three parts are introduced following: carrying out the Sanjiangyuan Ecological Protection and execution regulations of the Phase II Program; further strengthening scientific and technological investment and research; constructing a Sanjiangyuan Protection System involving participation of the entire society.

# CONTENTS

Preface / 085

## I  General Report

**G.1** Natural and Socio-Economic Characteristics of
the Sanjiangyuan Region / 089
    Section 1  Geographical Position / 089
    Section 2  Topography and Terrain Features / 093
    Section 3  Climate / 094
    Section 4  Surface Cover Features / 095
    Section 5  Species and Biological Communities / 096
    Section 6  Social and Economic Development / 097

## II  Subject Reports

**G.2** Hydrological Conditions and Water Resources / 099
    Section 1  Precipitation Characteristics / 099
    Section 2  River System / 107

| | Section 3 | Wetland Resources | / 131 |
| --- | --- | --- | --- |
| | Section 4 | Glacier Snow Mountain | / 135 |
| | Section 5 | Groundwater Resources | / 138 |

**G.3 The Course and Achievements of the Sanjiangyuan Ecological Environment Protection** / 140

| | Section 1 | Establishment of the Sanjiangyuan Comprehensive Pilot Zone and National Park | / 140 |
| --- | --- | --- | --- |
| | Section 2 | Obvious Implementation Effect of the General Plan for the Nature Reserve | / 144 |
| | Section 3 | Phase II Plan of the Sanjiangyuan Comprehensive Ecological Protection Pilot Zone | / 149 |
| | Section 4 | Action of Community in the Society | / 156 |

**G.4 The Outlook of the Sanjiangyuan Ecological and Environmental Protection** / 163

| | Section 1 | Challenges and Problems of Protecting Sanjiangyuan | / 164 |
| --- | --- | --- | --- |
| | Section 2 | Carrying out the Sanjiangyuan Ecological Protection and Execution Regulations of the Phase II Program | / 166 |
| | Section 3 | Further Strengthening Sicentific and Technological Investment and Research | / 169 |
| | Section 4 | Constructing a Sanjiangyuan Protection System Involving Participation of the Entire Society | / 171 |

**References** / 173

**Acknowledgements** / 174

# Preface

"How could the water be so clear and cool? Because fresh water comes from the source." This is extracted from the poem written by Zhu Xi, the famous poet in the Song Dynasty, and is also suitable to describe the headwaters of the Yangtze River, the Yellow River and the Lancang River, namely the Sanjiangyuan Region, in Qinghai Province. The Sanjiangyuan Region provides 40 billion cubic meters of water to the middle and lower reaches of these three great rivers, including 1.3% of the Yangtze River, 38% of the Yellow River and 15% of the Lancang River. Therefore, the region is significant for the provision of the three rivers, which is nicknamed as the Source of the Rivers, the China Water Tower and the Asian Water Tower. It is also the biggest area with the most variety of the high-plateau organisms in the world as well as the most sensitive region and the starting point of climate change in Asia, the Northern Hemisphere and the whole world. It is one of the most important sections of the ecological security screen in China as well as in the world. It is not only the source of some great rivers in China and Southeast Asia but also the origin of billions of lives and the power of the social and economic development for the countries alongside those rivers. Therefore, it is very important to protect this region.

The Sanjiangyuan Region, nicknamed as the ecological virgin land, used to be the high-plateau grassy marshland with rich grass, many lakes and a variety of wild animals. It is the region that has been steadily providing water for the Yangtze River and the Yellow River, and nurturing the Chinese civilization till today. Since the end of the $19^{th}$ century and the beginning of the $20^{th}$ century, climate change has been causing the warming of the globe and the shrinking of the glaciers and snow mountains, which cut the water provision to the high land lakes and wetlands. Many lakes and wetlands started to shrink; some even dried.

The ecology has turned out to be more fragile. With the growth in population and the increases in human activities, the region is exposed to ecological deterioration. People from all walks of life have noticed the change. The national water resource security, ecological security and sustainable development are threatened due to the damage of the plants and ecological system in the region, as well as the declination of water conservation capacity.

The central government pays significant attention to the protection and construction of the Sanjiangyuan Region by providing important guidance. In 2016, President Xi Jinping stressed on his trip to Qinghai that possesses special ecological position. Qinghai must shoulder the great responsibility of protecting the Sanjiangyuan Region and the China Water Tower. The manual remedy and the natural recovery must be combined when the protection is prioritized. The plans must be based on the local situation to strengthen the function of the national ecological security. The work, such as the ecological project, energy conservation and emission reduction, environmental pollution remediation, construction of beautiful cities and countrysides, reinforcement of the building of the natural reserves, realization of the trial of the Sanjiangyuan National Park, ecological protection around the Qinghai Lake, remediation of the desertification, building of the frigid grassland, returning the grain plots/graze land to forestry/grassland and the protective forest planting in North China and the Northeast China, must be comprehensively pushed forward so as to make sure the middle and lower reaches of the great rivers are clean and clear.

In 1999, the China Expedition Association made an overall inspection on the Lancang River and proposed "developing the West of China and Protecting the Sanjiangyuan Region", which was echoed by the State Forestry Bureau, Qinghai Provincial Government, the Chinese Science Association and the Chinese Academy of Sciences. In March 2000, the forum of the feasibility of the Sanjiangyuan Natural Reserve was jointly hosted by the State Forestry Bureau, the Chinese Academy of Sciences and the Qinghai Provincial Government. In May 2000, the Sanjiangyuan Natural Reserve at the provincial level was approved by the local government. In 2005, the *General Plans of the Protection and Construction of the Ecology at the Sanjiangyuan Natural Reserve* was

# Preface

approved by the State Council and the project which is praised as the Chinese No. 1 ecological project in the new century was unveiled then. In November 2011, the State Council decided to establish the comprehensive testing area of the State ecological protection and approved the corresponding general plans. The testing area was the first try in China with the aim of building the Sanjiangyuan Region into an advanced and demonstrative area of Eco-civilization, developed economy, wealthy life and harmonious society. In 2014, the Phase II project of the testing area was started.

The protection of the Sanjiangyuan Region is a significant, challenging and heavy-duty task. It concerns not only the profits of the people living in the region but also the sustainable development of the country, the Eco-construction and the realization of the China Dream; therefore, the protection of the area is closely related to every Chinese and is the task of every Chinese. In 2012, the Sanjiangyuan Protection Fund was established by the Qinghai provincial government in order to get more support and care from the society and build a bigger platform for people to participate in the protection activities. Since then, related fund-raising has been organized many times, and the donation has been from both home and abroad. The fund sponsors the resources of the Sanjiangyuan Region, ecological protection projects, related researches, technological development and publicity of the region, academic exchange, international exchange and cooperation.

Under the guidance of the *General Plans of the Protection and Construction of the Ecology at the Sanjiangyuan Natural Reserve* and efforts from the government and people, the ecology of the region has been improved greatly. The Ecological system structure is partially improved; the deterioration of the grassland is basically curbed; the contradiction between the grassland and grazing is lessened; the ecological function of the wetland is improved step by step; the lakes are enlarged and the water provision ability is strengthened. However, the ecology in the region still has its problems, such as the worsening of the grassland, the snow line rising and the shrinking of the glaciers. There is a long way to go for the ecological protection and the rescue of the China Water Tower and it needs additional scientific measures as well as participation from more stakeholders.

It is the better knowledge of the region that brings better protection and development to it. The searching for the source land of the three great rivers has never stopped since ancient times. In the *Classic of Mountains and Rivers*, there are words like "the source of the great rivers lies in the Kunlun Mountain"; In the *Searching of the River Sources* written by Xu Xiake, the famous geologist in the Ming Dynasty, Jinsha River was regarded as the source of the Yangtze River. In 1978, the scientific expedition of the Yangtze River decided that the Yangtze River had three sources: Ulan Moron, Nanyuan Dangqu and Xiyuan Chuma'er River. In 2008, the research and search of the great river sources were still under process with the help of satellite positioning and remote sensing technology. This searching is the manifestation of the Chinese spirit of seeking for the truth.

*The Green Book on the Sanjiangyuan Region*, with the encouragement of such spirit and the help of the modern technology, aims at creating a better understanding of the current situation of the region for those who care about the development of the area. The Book also hopes to be the reference to some governmental policy making and scientific researches. Under the advocacy and instruction of Professor Wang Guang-qian, Academician of the Chinese Academy of Sciences, the Green Book is composed with the help of the Sanjiangyuan Ecological Protection Fund. According to the plan, the Book consists of five volumes, which will be on the water resources, the land use, the traditional culture, the ecological protection and the economic development, respectively. The first volume talks mainly about the water resources in the region. The data includes those that already exist and new information achieved through the technology of satellite precipitation inversion and Tsinghua Hydro30, a HR worldwide river net that is established by Tsinghua University.

*The Green Book* is far from perfect; therefore, more valuable suggestions will be greatly appreciated.

The Editorial Committee of
*The Green Book on the Sanjiangyuan Region*
July, 2017

# General Report

## G.1 Natural and Socio-Economic Characteristics of the Sanjiangyuan Region

### Section 1  Geographical Position

The Sanjiangyuan Region is the hinterland of the Qinghai-Tibet Plateau and lies in the south of Qinghai, a province in the west of China. However, the fundamental connotation and the geographical range of this region is different due to different divisions. The area mainly includes the water basins of the three great rivers, the Sanjiangyuan National Reserve, the comprehensive testing area of the Sanjiangyuan State Ecological Protection and the Sanjiangyuan National Park. As there has been no consensus on the boundary of the region to date, there are different descriptions or reports on the area, which cause confusion to the readers. Therefore, this chapter attempts to distinguish between the major different connotations of the region.

#### 1. The Headwater Basins of the Three Great Rivers

According to the natural river basins, the Sanjiangyuan Region is the source land of the Yangtze River, the Yellow River and the Lancang River with a basin area of 312,600 km². The basin area of the Yellow River is 122,000 km² (the area refers to the far western part of the Tangnaihai Hydrologic Station – *Annals of China Water Conservation Statistics*), which accounts for 39% of the total area of the Sanjiangyuan Region. The basin area of the Yangtze River is 137,700 km² (it refers to the far western part of the Zhimenda Hydrologic Station), which accounts for 44.40% of the total area of the Sanjiangyuan Region. The basin area of the Lancang River is 52,900 km² (it refers to the far western part of the Qamdo Hydrologic Station), which accounts for 16.90% of the total area of the Sanjiangyuan Region. The areas of the basins of these three great rivers within the Sanjiangyuan Region are show in Figure 1.1.

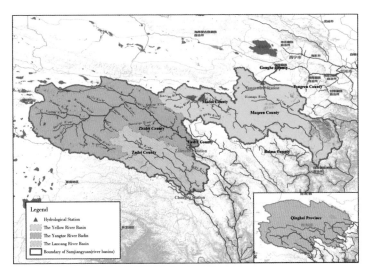

**Figure 1.1  The scope of the Sanjiangyuan Region defined according to the natural water basins (312,600 km²)**

## 2. The Sanjiangyuan Nature Reserve

In May 2000, the Sanjiangyuan Nature Reserve was approved by the Qinghai Provincial Government in order to protect the regional ecological environment. In January 2003, the State Council approved the Sanjiangyuan Nature Reserve to be a national one. In January 2005, the State Council

## Natural and Socio-Economic Characteristics of the Sanjiangyuan Region

approved *The General Plans for the Ecological Protection & Construction at the Sanjiangyuan Region in the Qinghai Province*. The protection area was 152,300 km² out of the total area of 312,600 km² and consists of 69 counties in the Yushu Prefecture, the Golog Prefecture and the Huangnan Prefecture. The reserve is defined according to its function as the core area, service area and trial area with 20.5%, 25.8% and 53.7%, respectively.

### 3. The Sanjiangyuan National Ecological Protection Comprehensive Testing Area

In 2011, the 181st executive meeting of the State Council approved the *General Plan for the Sanjiangyuan National Ecological Protection Comprehensive Testing Area*. An area of 395,000 km² was defined as the comprehensive testing area, including 21 counties in the Yushu Prefecture, the Golog Prefecture, the Hainan Prefecture, the Huangnan Prefecture and the Tanggula County in Golmud City as well as 158 villages and towns. The *General Plan* divides the testing area into the key protection area, ordinary protection area and transferring area, with areas of 50.1%, 47.9% and 2%, respectively. The comprehensive testing area is shown in Figure 1.2.

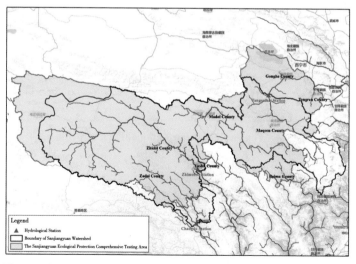

Figure 1.2 The Sanjiangyuan National Ecological Protection Comprehensive Testing Area (395,000 km²)

## 4. The Sanjiangyuan National Park

The planned area of the Sanjiangyuan National Park includes the source of the three great rivers, the upgraded Kekexili (Hoh Xil) National Nature Reserve and five districts of the Sanjiangyuan Nature Reserve to form the layout of a park with three districts, namely the Yangtze River Park, the Yellow River Park and the Lancang River Park with a total area of 123,100 km², which accounts for 31.2% of the total testing area. Within this, glaciers cover 833.4 km²; rivers, lakes and wetlands cover 298,428 km²; grassland accounts for 868,322 km²; and forests cover 495.2 km². Please refer Figure 1.3 for details.

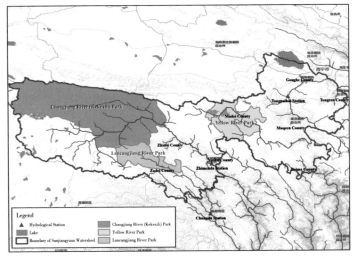

Figure 1.3 The Sanjiangyuan National Park (123,100 km²)

The Yangtze River Park (the Hoh Xil) lies in the Zhidoi County and the Qumarlêb County of the Yushu Prefecture, including the Hoh Xil National Nature Reserve and the Suojia-Quma? River Nature Reserve in the Sanjiangyuan National Nature Reserve, with an area of 90,300 km² and consisting of 15 administrative villages.

The Yellow River Park lies in the center of the Sanjiangyuan Region.

The park holds two natural sub-protection areas—the Gyaring-Ngoring Lake and the Xingxing Sea, with an area of 191,000 km$^2$ and consisting of 19 administrative villages.

The Lancang River Park is in the Zadoi County of the Yushu Prefecture, including the Guozongmucha and the Angsai Nature Reserves of the Sanjiangyuan National Nature Reserve. The park has an area of 13,700 km$^2$ and consists of 19 administrative villages.

This Green Book focuses on research on the Sanjiangyuan National Ecological Protection Comprehensive Testing Area. Therefore, except where specifically noted, any mention hereafter of the Sanjiangyuan Region refers only to the Sanjiangyuan National Ecological Protection Comprehensive Testing Area.

## Section 2  Topography and Terrain Features

The Sanjiangyuan stands in the hinterland of the Qinghai-Tibetan Plateau, the youngest plateau in the world. It used to be a sea 200 million years ago. Then, it kept rising to form a continent. Several millions of years ago, the continent rose greatly to form the Qinghai-Tibetan Plateau, and the rising is still on-going. The topography of the Sanjiangyuan mainly consists of mountains and valleys, whose landscape is complex. The altitude is between 3,335 and 6,564 meters, with an average level of 4,000 meters. The basic frame is mountain ranging from 4,000 to 5,800 meters high. The main mountains are the east Kunlun Mountains, the Amne Machin Maqên, the Bayankela (Bayan Har) Mountains and the Tanggula Mountains.

Because of their special geographic positions and different altitudes, the Yangtze River Source Park and the Lancang River Source Park enjoy glacier landforms, hills, mountains, rivers, lakes and watersheds. The Yellow River Source Park has a relatively lower altitude, and it mainly consists of hilly mountains, lakes and river basins.

In the central, western and northern parts of the region, the landscape is quite flat with beaches, alpine meadows, and big area of swamps; the southeastern part consists of high mountains and valleys. Much of the area has slopes of 30 degrees or above with broken and steep terrains, and the altitude difference is more than 1,000 meters. The landscape has primitive forest distributed throughout. In the northeast, the Yellow River gradually lowers its way from an altitude of 3,306 meters (the Tangnaihai Hydrologic Station in the Xinghai County) to 1,960 meters (in the Jainca County), and so the landscape is quite flat, with a combination of valleys, basins, wetlands and terraces.

## Section 3　Climate

The climate of the Sanjiangyuan Region is mountain climate of the Qinghai-Tibetan Plateau, a typical continental plateau climate which features alternative warm seasons and cold ones, distinctive dry and humid days, rain heat synchronization, small annual temperature difference, big daily temperature difference, long duration of sunshine, strong solar radiation, short vegetation period and very short or almost no frost-free period.

The average annual temperature is in the range of -5.6-3.8 ℃. The average temperature of the coldest month (January) is in the range of -13.8-6.6℃ and the lowest temperature is about -48 ℃. The average temperature of the hottest month (July) is in the range of 6.4-13.2 ℃ and the highest temperature is about 28 ℃. According to the latest surface meteorological data collected by the China Meteorological Administration, the temperature of this area has been rising during the past several years. The average Thiessen figure of the annual average temperature collected from every meteorological station was -1.52 ℃ during the period 1981-2000, while it was -0.45 ℃ during the period 2010-2013. The average climate change rate was 0.32℃ /10 years from 1981 to 2013. Before 2000, the average altitude with the temperature above 0℃ was 3,646 meters, while it raised to 3,968 meters after 2000, indicating that the average altitude with an average temperature of 0 ℃ rises as the climate changes.

The annual average precipitation of this area is 406.6 mm, of which 85.3% occurs from May to July. About 55%-66% rain falls at night, and the annual evaporation capacity reaches 730-1700 mm. There are 2300-2900 hours of sunshine, with a solar radiation of 5,658-6,469 MJ/m2 in this area. The area suffers from gales of force eight and above from 3.9 to 110 days in a whole year. The air contains 60%-70% as much oxygen as it does at the sea level. The cold season is controlled by the cold anticyclone of the Qinghai-Tibetan Plateau for seven months in a year, with the characteristics of low heat, less precipitation and strong wind. The warm season is influenced by the southwest monsoon, and so it features hot barometric pressure, more precipitation, humid weather and frequent night rainfalls.

## Section 4   Surface Cover Features

The Sanjiangyuan State Ecological Protection Comprehensive Testing Area covers 395,000 km$^2$. The grassland of the region shares 280,200 km$^2$, taking 70.94%; the forestry covers 22,300 km$^2$, taking 5.65%; the farm land is 1,178 km$^2$, only taking 0.30%; the land for other agricultural purpose is 431 km$^2$, taking 0.11%; the villages (towns) and the mining places covers 301 km$^2$, taking 0.08%, waters and the water conservancy facilities share 5000 km$^2$, taking 1.27%; the land traffic is 127 km$^2$, taking 0.03%; and the area of undeveloped land and land for other uses is 85,400 km$^2$ (including the forestation areas), taking 21.62%. Furthermore, according to the data collected through the inversion of remote sense there are 22,800 km$^2$ bare land in the region, taking 5.95%; the glaciers and firn covers 3,073 km 2, taking 0.80%. Due to the special climate, the land surface is greatly influenced by the coldness and freezing, the soil develops slowly and the soil-forming period is short. Therefore, the soil is barren, young, coarse and sandy (mainly fine snads, cuttings and gravels), the soil is also featured with poor water conservancy capacity and high erosion.

The soil can be divided into 15 types and 29 subdivisions. Due to the mountainous landscape and different altitudes, the soil is vertically distributed from high places to the lower points, which consists of alpine desert soil, alpine meadow soil, alpine grassland soil, mountain meadow soil, gray cinnamonic soil, chestnut soil and mountain forest soil. The alpine meadow soil is the majority and it spans from the altitude of 3,500m to 4,800m. The swamp meadow is fairly developed and the frozen soil covers a rather big area.

## Section 5    Species and Biological Communities

The Sanjiangyuan Region raises very special species and biological communities thanks to its alpine characteristics, thus it is a valuable gene bank of plateau species.

The vascular plants have 87 families, 471 categories and 2,308 types, taking 8% of the plants types across China, among which, 8.5% belongs to seed plants; 422 (89%) categories of herbaceous plants consist of 2,125 (92.07%) types; 41 (8.7%) shrub categories make 144 (6.24%) types; 11 (2.3%) arbor categories make 39 (1.69%) types. The majority plants are herbaceous and the vegetation, they mainly consist of coniferous forest, broadleaf forest, conifer-broadleaf forest, shrubs, meadows, grassland, swamps & aquatic vegetation, cushion vegetation and sparse vegetation. The region enjoys Picea brachytyla var, Complanata, red flower meconopsis, and Chinese caterpillar fungus that fall into the Category II of the National Protected Plants. 31 types of orchids that are listed in the Category II of the International Trade Convention and 34 types plants that are Qinghai Provincial protected ones growing in the area. In this region, more plants, fruits and herbs are developed to suit the alpine ecological environment.

Many wild animals live in this region. There are 8 orders of beasts, 20 families and 85 species, taking 16.8% of that in the whole country. There are 16 orders, 41 families and 238 species (including 263 subdivisions) of birds in this region, taking 19% of that in the whole country; there are 7 orders, 13 families and 48 species of amphibious reptile in this region. There are 6 families of fish, making 40 species, and more than 40% of them are unique in China. There are 69 species of animals that are key national protected ones, such as Tibetan antelopes, wild yak, snow leopard, argali, and goa. There are 32 species of provincial protected animals, such as mustela putorius and corsac fox. Furthermore, the unique species have contributed a lot to the development of the local Tibetan and Chinese herbs, agricultural and husbandry products processing and domestication and cultivation of certain wild animals.

## Section 6　Social and Economic Development

In 2015, there were 1.3281 million people in the Sanjiangyuan Region, accounting for 23.1% of the total population of Qinghai. Among them 279,800 are non-agricultural residents, accounting for 21.1% of the regional population while 1.0483 million farmers and herdsmen take the percentage of 78.9% of the regional population. There are Han, Tibetan, Mongolian and Hui people in the region, among which Tibetan accounts for 80%.

Although there are less people (less than 4 people/km$^2$) in the area, the ecology is still challenged by human activities. According to the statistics from the middle of the Qing Dynasty to the beginning of the Republic of China (100 years), the population of this area has increased by 75%, with an annual increasing rate of 5.6‰. The population increased two times every 125 years in this period. During the period of the Republic of China (1912-1949), the population has increased by 90% with an annual increasing rate of 18.6‰. That turns two times every 38 years. Since the founding of the People's Republic of China to the sixth census of population in 2010, the population has increased six times with an annual increasing rate of 20.6‰. That turns two times every 21 years.

The traditional industry in the region is grassland livestock husbandry, it has very weak foundation from a social-economic point of view. Before the founding of the People's Republic of China there was little income from any other industries except handicraft and ethnic supplies industry in the area. Since 1949, there are industries, such as coal, power, wood processing, construction materials production, gold mining, agricultural and livestock product processing, developed in this region. In recent years, some tertiary industries, such as ecological tourism, adventure tourism, custom tourism and religious pilgrimage tour, have been developed.

In 2015, the total output value of this region is 30.916 billion RMB, accounting for 12.8% of the provincial gross. Among them the primary industry is 8.243 billion RMB, the secondary industry is 13.254 billion RMB and the tertiary one contributes 9.419 billion RMB. The public income (2.255 billion RMB) is as 13 times as the expenditure (31.302 billion RMB). The per capita disposable income of the local residents was 12.4 thousand RMB, only accounting for 56.2% of that of the whole country. Generally speaking, the region is underdeveloped with eight key poverty alleviation counties at a national level, eight such counties at a provincial level and many poor people living in this region.

# Subject Reports

# G.2
# Hydrological Conditions and Water Resources

## Section 1  Precipitation Characteristics

Precipitation is the main resource for surface water. Its spatial, temporal distribution and variation directly determine a region's humidity and dryness, as well as its amount of water resources and so on. Sanjiangyuan is located in the hinterland of the plateau, and it has the feature of high altitude, complicated terrain, high evaporation and other characteristics. The region's precipitation characteristic is also different from other regions. Therefore, studying the spatial and temporal distribution as well as variable trend of precipitation in Sanjiangyuan is very important to understand the hydrology and water resources in this region.

Due to the small population of the region and large sparsely-populated area, there are only 23 long-term monitoring rainfall stations on the nearly 400,000 square kilometers land. So the coverage density of station is very low, thus ground monitoring data cannot really reflect regional precipitation distribution. With the development of satellite remote sensing technology, satellite precipitation data products can effectively compensate the shortcomings of the monitoring site on earth. There are two widely used satellite precipitation data products existing now. One is the CMORPH data from the National Oceanic and Atmospheric Administration (NOAA), and the other is the TRMM data released by the National Aeronautics and Space Administration (NASA) and the Japan Aerospace Exploration Agency (JAXA). The results show that the data collected by these two products is almost consistent in the Qinghai-Tibetan Plateau, while the spatial resolution of the COMRPH data is higher (maximum spatial resolution of the CMORPH data can be 8 kilometer, i.e. there is a point in every 8 kilometers, and spatial resolution of the TRMM data is only 0.25 degrees, about every 20 kilometers per point), so the CMORPH data has more advantages in this region. The CMORPH data collection began in 1998. In the first two years of initial operation, the satellite data acquisition and processing methods were not yet mature, so the accuracy of data was relatively low. So the data analysis on the precipitation in the the Sanjiangyuan area is made by using 2000 -2015 CMORPH data. With more frequent use of the data by more scientific research institutions, the technology has gained high accuracy in data analysis, especially in the study of a wide range of rainfall distribution.

## I  Spatial Distribution

Based on the data of the CMORPH satellite precipitation, the spatial distribution of average annual precipitation in the Sanjiangyuan area from 2000 to 2015 is shown in Figure 2.1. On the whole, the spatial distribution of precipitation in the region shows the pattern of gradual reduction from the southeast to the northwest, and the change is of well distribution in space. The annual precipitation in the western and northern regions is generally less than 300 millimeters, these regions belong to arid and semi-arid area. The precipitation in the central region is between 300 and 700 millimeters. The annual precipitation in the eastern and southern regions can reach more than

# Hydrological Conditions and Water Resources

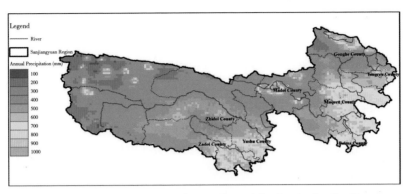

Figure 2.1 Spatial distribution of annual precipitation in the Sanjiangyuan Area (2000–2015)

700 millimeters, and these regions belong to semi-humid area. From 2000 to 2015, the average annual precipitation in the Sanjiangyuan area is 406.6 millimeters, and the total annual precipitation resource is about 160.6 billion cubic meters. The proportions of precipitation in the Yangtze River and the Yellow River are very close, both with 36.5%. The precipitation resources of the Lancang River and the inland drainage area are 14.1% and 12.8%, respectively.

It should be noted that due to the systematic deviation of inland water precipitation from the CMRPH satellite, the estimated amount of precipitation in the lake areas of the Sanjiangyuan region is significantly large. As can be seen in Figure 2.1, the precipitation goes up suddenly in the Qinghai Lake, the Ngoring Lake (the Eling Lake), the Gyaring Lake (the Zhaling Lake) and the inland lakes of the Changtang (the Qiangtang) Plateau, which are unusually higher than the areas outside lakes. This phenomenon also occurs in other regions, as foreign scholars have pointed out that the CMORPH and the TRMM satellite precipitation data have been overvalued significantly in the inland lake area in northeastern United States. Although the problem of satellite precipitation data will increase the error of precipitation estimation in Sanjiangyuan, it has a little effect for us on the study of the law of precipitation spatial distribution.

The Yangtze River Basin in the Sanjiangyuan occupied an area of 158,000 square kilometres. In addition to the mainstream of the Jinsha River originated from the Sanjiangyuan, the Yangtze River tributaries—the Min

River and the Yalong River— are also originated from Sanjiangyuan. The annual precipitation is 366.1 millimeters and the average annual precipitation resource is 58.67 billion cubic meters. The upstream of the Jinsha River has less precipitation in the western part of Sanjiangyuan, with an annual precipitation between 150 and 400 millimeters, and the annual precipitation in the upstream of the Min River and the Yalong River is between 500 and 750 millimeters.

The Yellow River Basin occupies an area of 119,000 square kilometers of the Sanjiangyuan area, with an average annual rainfall of 487.3 millimeters and an average annual precipitation resource of 58.66 billion cubic meters. The precipitation in this area is relatively high in the east, and low in the west. In the eastern of the Tongren County, the Zêkog (the Zeku) County, the Tongde County, the Maqên (the Maqin) County, the Henan Mongol Autonomous County, the Gadê (the Gande) County and the Jigzhi (the Jiuzhi) County, the precipitation is more than 500 millimeters. In the Yellow River Basin, above the Gadê (the Gande) County and the Longyangxia Dam area, the precipitation is generally less than 500 millimeters.

The Lancang River Basin occupies an area of 37,000 square kilometers of the Sanjiangyuan area. Its annual precipitation is 605 millimeters and the average annual precipitation resource is 22.65 billion cubic meters. Since the precipitation along the river rises from 300 millimeters to 800 millimeters, the area is more abundant in rainfall in the Sanjiangyuan area.

Other areas in Sanjiangyuan belong to the inland drainage area, covering an area of 81,000 square kilometers. Although the river stream does not flow to the Sanjiang region directly, it still has an important impact on local ecosystem. The average annual precipitation in the interior drainage area is 268.3 millimeters, the total annual precipitation is 20.64 billion cubic meters.

## II  Temporal Distribution

The precipitation in Sanjiangyuan has obvious seasonal characteristic. The process of precipitation change in each month from 2000 to 2015 is shown in Figure 2.2. Precipitation from May to September accounts for 85.3% of the annual precipitation, and precipitation from June to August accounts for 59.3%

of the annual precipitation. The precipitation in July is the highest of a year. The uneven distribution of precipitation aggravate the drought situation in Sanjiangyuan, which brings challenges to water resources management locally.

The monthly change of the precipitation in Sanjiangyuan is shown in Figure 2.3. It can be seen that the seasonal variation of precipitation is obvious. The Lancang River has the highest precipitation, then followed by the Yellow River, the Yangtze River and the inland lakes which share close precipitation. Taking the different areas of different drainage areas in Sanjiangyuan into account, the differences of precipitation do not represent the differences of the total amount of water resources of each river basin.

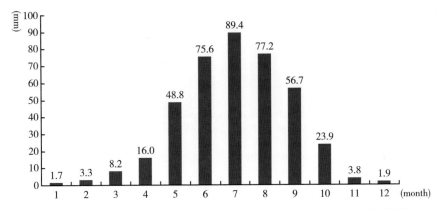

**Figure 2.2 The average monthly precipitation in Sanjianyuan (2000–2015)**

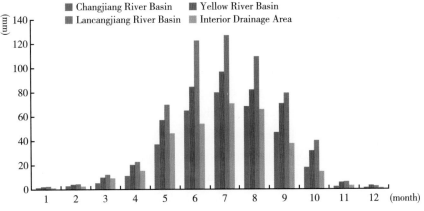

**Figure 2.3 The average monthly precipitation in each basin of Sanjiangyuan (2000–2015)**

The change of precipitation in the Sanjiangyuan area from 2000 to 2015 is shown in Figure 2.4, the annual precipitation fluctuates between 370 millimeters and 430 millimetres whose difference is not big. The data shows that the annual precipitation in the Sanjiangyuan area increased significantly, about 1.69 millimetres per year. The tendency is shown by the dotted line in Figure 2.4.

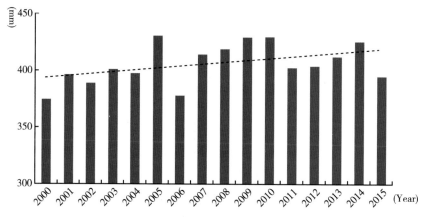

**Figure 2.4   Precipitation change in Sanjiangyuan from 2000 to 2015**

Due to short observation period of satellite precipitation, the rainfall data of stations is used to extend the satellite precipitation sequence in order to observe the long-term variation of precipitation in the Sanjiangyuan region. Using the Thiessen Polygon Method, the precipitation data of 23 rainfall stations in the region is taken as an average of the whole area, thus the precipitation change from 1960 to 2015 in the region is obtained (See Figure 2.5). The results show that the average annual precipitation in the region fluctuates near 400 millimetres, the change of precipitation is not obvious before 1994, but after 1994 it starts to rise. The average precipitation from 2000 to 2015 is 434.3 millimetres, which is slightly higher than 406.6mm provided by the satellite data, and in most of the years the precipitation recorded by stations is higher than that recorded by satellite. Due to the sparse distribution of meteorological stations in the Sanjiangyuan region and lack of

Hydrological Conditions and Water Resources

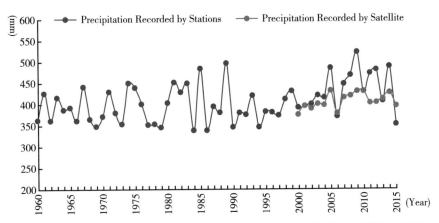

**Figure 2.5  Precipitation change in the Sanjiangyuan region from 1960 to 2015**

spatial fineness, the accuracy of spatial characteristics of precipitation is not good enough compared with satellite precipitation data, therefore it can be only taken as a reference.

Based on precipitation recorded by satellite, the interannual variability of precipitation resources in each river basin is shown in Figure 2.6. It can be seen that the changing rule of precipitation in different drainage areas is not synchronized. In addition, the changing trend is relatively independent from each other.

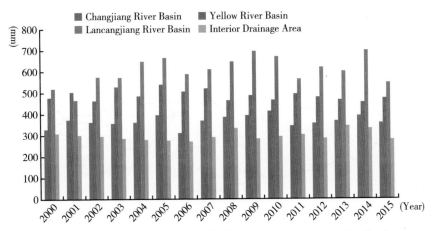

**Figure 2.6  The change of total precipitation of each river basin in Sanjiangyuan from 2000 to 2015**

105

## III  Satellite Data Validation

In order to verify the reliability of satellite data, the data provided by rainfall stations is used for comparison. The comparison shows that the data from rainfall stations is relatively reliable, thus can be treated as the true value of precipitation. The distribution of 23 rainfall stations in the Sanjiangyuan region is shown in Figure 2.7, where most of the stations are located in the eastern and central areas of the Sanjiangyuan region. By making comparative analysis of annual precipitation provided by stations and satellite, it is found that the average error rate of annual precipitation provided by satellite is between 10% and 20%, except for a few places with a high rate of error. Overall, the higher the altitude in the area is, the relatively lower accuracy in satellite precipitation data shows. However, due to the low-frequent precipitation in the western high altitude areas, the contribution of this part to the total precipitation is small, so the effect of the error is relatively small. It can be seen from Figure 2.8 that, in addition to the high error rate (25%) of the Ulan Moron (the Tuotuohe) station, the satellite data of the Qumarlêb (the Qumalai) station, the Darlag (the Dari) station, the Xinghai station has high conformity to the data measured by stations, so the deviation is relatively

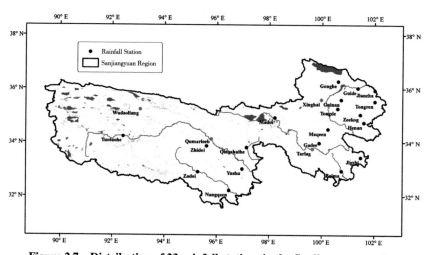

**Figure 2.7  Distribution of 23 rainfall stations in the Sanjiangyuan region**

# Hydrological Conditions and Water Resources

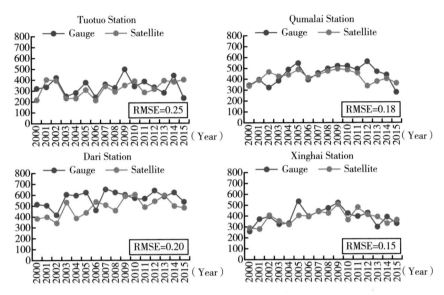

**Figure 2.8** Comparison of recorded precipitation and satellite precipitation data in meteorological stations

small. Due to high elevation and complicated topography, for the areas 20 km near the Ulan Moron (the Tuotuohe) station, the data measured by the station is not of representation. While the other three meteorological stations are with relative flat location, so the data obtained there is relatively reliable.

Compared with the data from rainfall stations, satellite precipitation data has obvious advantages in spatial coverage. Especially for areas with few rainfall stations, satellite retrieval of precipitation can monitor the temporal and spatial distribution characteristics of precipitation in a larger range and help find new rules. So the satellite data is used in this paper for precipitation analysis.

## Section 2  River System

The Sanjiangyuan water system can be divided into outflow water system and internal water system. Outflow system includes the Yangtze River source river basin, the Yellow River source river basin, and the Lancang

River source river basin. The internal river system, includes the Changtang (the Qiangtang) Plateau water system, the Qaidam (the Chaidamu) Basin water system and the Qinghai Lake water system. The structure of the Sanjiangyuan water system is studied based on Tsinghua University's independent research and the data from high-resolution digital river network "Tsinghua Hydro30". Based on 30 meters of global DEM data, "Tsinghua Hydro30" is of the quality with high degree of fineness by using improved river network extraction method, it is accurate in positioning and rich in levels and so on. Due to the complicated terrain, dense river networks, and large areas without people, it is difficult to make field measurements for all rivers in the Sanjiangyuan region. "Tsinghua Hydro30" technology can accurately identify the region's river water distribution and structure, and it is of great significance to further implement the ecological protection measures. The distribution of the drainage area is shown in Figure 2.9. The outflow area of the three major river basin areas and the number of tributary are as shown in Table 2.1. According to the statistics, there are 1,416 rivers with river basin area above 50 square kilometers.

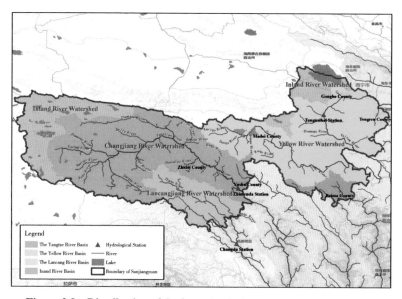

**Figure 2.9 Distribution of drainage basin in the Sanjiangyuan region**

# Hydrological Conditions and Water Resources

**Table 2.1 Basic information of three river basins in the Sanjiangyuan region**

(a) Drainage area

| Drainage | Area (ten thousand square kilometers) | The proportion of the total area of Sanjiangyuan | Length of the main stream | The proportion of the total length of the mainstream |
|---|---|---|---|---|
| The Yellow River | 11.9 | 30.1% | 1983 | 34.9% |
| The Yangtze River | 15.8 | 40.6% | 1206 | 19.1% |
| The Lancang River | 3.7 | 9.4% | 448 | 20.4% |

(b) Number of tributaries at all levels

| Drainage area | The number of tributaries | | | |
|---|---|---|---|---|
| | First class tributary | Second class tributary | Third class tributary | Fourth class tributary |
| The Yellow River | 126 | 338 | 157 | 8 |
| The Yangtze River | 109 | 274 | 162 | 30 |
| The Lancang River | 46 | 108 | 51 | 7 |

Note: The minimum river basin area is set about 50 square kilometers.

## I  Water System in the Yellow River Source Area

The Yellow River is originated from the Yueguzonglie basin in the north of the Bayan Har (the Bayankela)Mountains. The altitude of the source area is 4,724 meters. The river runs into Gansu from Sigou Gorge, this part is roughly in an "S" shape. From the river source to the Sanjiangyuan exit, The Yellow River goes through 1,983 kilometers, accounting for 34.9% of the total length of 5,687 kilometers. The Yellow River Basin has an area of 119, 000 square kilometers, accounting for 30.1% of the total area of the Sanjiangyuan region. There are 126 tributaries with a basin area of more than 50 square kilometers, of which there are 4 rivers (Qiemu Qu, Duo Qu, Re Qu, Qushi'an River) with a basin area of more than 5,000 square kilometers, 21 rivers with 1,000-5,000 square kilometers, and 45 rivers with 500-1,000 square kilometers. There are many Second-Class tributaries and even below the Second Class tributaries. The main tributaries of the Yellow River source area are shown in Figure 2.10. Within the Sanjiangyuan region, those rivers with large basin

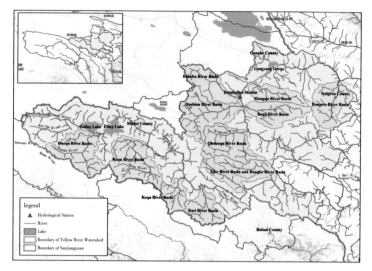

Figure 2.10　Main tributaries of the Yellow River

areas and volume of runoff includes the Duo Qu basin, the Re Qu basin, the Ke Qu basin, the Darlag (the Dari) River basin, the Xike Qu and the Dongke Qu basin, the Qiemu Qu basin, the Qushi'an basin, the Daheba River basin, the Ba Qu basin, the Mang Qu upstream basin, the Longwu River basin and so on.

The Duo Qu is originated from the north of the Chindu (the Chengduo) County, which is located in the south of the Gyaring (the Zhaling) Lake, flowing into the Ngoring (the Eling) Lake in the Madoi (the Maduo) County. The basin area is 5,905.14 square kilometers, and the elevation is 4400 - 4600 meters (from Google Earth, the same below), declined from southwest to northeast. The Duo Qu basin is of a cold temperate continental monsoon climate with an average annual rainfall of 300 to 400 millimeters (from the hydrological annual book, the same as below). The average annual runoff is 420 million cubic meters, with the natural drop of 536 meters (from the *Chinese Water Dictionary*, the same below). The midstream and downstream river of the Duo Qu on both sides have a lot swamps and grass, and the animal husbandry there is developed.

The Re Qu is originated from the south of the Madoi (the Maduo) County, and the north of The Darlag (the Dari) County. The two tributaries of it is Re Qu in the eastern part and the Hei River in western part, which

## Hydrological Conditions and Water Resources

flow into the Yellow River after convergence.

Most of the drainage area is in the Madoi (the Maduo) County. The elevation is 4200 - 4800 meters, the terrain is high in southwest and low in northeast. The basin area is 6,702.00 square kilometers. The drainage area belongs to alpine grassland climate, with the average annual rainfall of 400 to 500 millimeters. The average annual runoff of the Re Qu is 680 million cubic meters, with a natural drop of 585 meters. Along the Re Qu, grassland is widely distributed. With abundant water and grass, animal husbandry is the dominating economy for the local people.

The Ke Qu is originated from the drainage area of the Darlag (the Dari) County in Qinghai Province and the Sêrxü (the Shiqu) County in Sichuan Province, flows from south to north into the Yellow River. The drainage basin area is 2,455.91 square kilometers with an elevation of 4300-4900 kilometers. The Ke Qu valley basin has a cold and humid climate, with no obvious four seasons. The average annual rainfall is 500 to 600 millimeters. The grassland covers wide areas, accounting for more than 90% of the land. The Darlag (the Dari) County is not only an important distribution center for goods, Chinese and Tibetan herbs, but also an important transportation hub for the Golog (the Guoluo) Tibetan Autonomous Prefecture.

The Darile Qu is originated from the middle of the Darlag (the Dari) County where the most water comes from, and the northwest of the Banma County. The two main tributaries are the Darile Qu and the Du Qu, which flow from south to north into the Yellow River after convergence. It has a drainage area of 3,400.61 square meters, and an elevation of 4100-4800 meters. The Darile Qu belongs to cold and humid climate, and the average annual rainfall is 600-700 millimeters. The average annual flow of the Dari River is 490 million cubic meters, and the natural drop is 116 meters. The ethnic groups in the valley basin are mainly Tibetan, as well as a few Han, Mongolian and Hui.

The Xike Qu and the Dongke Qu are adjacent, and both are located in the territory of the Gadê (the Gande) County. They run from northwest to southeast into the Yellow River, respectively. The Xike Qu basin area is 2,584.38 square kilometers, and the Dongke Qu basin area is 3,477.37 square kilometers with An elevation of 3800-4600 meters. The Xike Qu and the Dongke Qu belong to plateau continental semi-humid climate, with the average annual rainfall of 600-700 millimeters. The ethnic groups in the valley basin are mainly Tibetan, as well as a few Han, Hui and Tujia.

The Qiemu Qu is originated from the Zhuan Moutain- Maqinggangri Mountain in the Maqên (the Maqin) County, it flows from west to east into the Yellow River, and the valley basin area is 5,610.32 square kilometers, with an elevation of 3700-5400 meters.

The peak of Zhuan Mountain-Maqinggangri Mountain is 6,282 meters. The Qiemu Qu belongs to the continental humid climate with an average annual rainfall of 500 -600 millimeters. The average annual runoff of Qiemu Qu is 250 million cubic meters/second, and the natural drop is 1,830 meters. The Maqên (the Maqin) County is rich in grassland resources, and its animal husbandry is developed.

The Qushi'an River is originated from the border of the Madoi (the Maduo) County, the Maqên (the Maqin) County and the Xinghai County, it runs through the Xinghai County from west to east into the Yellow River. It has a valley basin area of 6,593.72 square kilometers. The terrain is high in the southwest and low in the east. The elevation is 3,100-5,200 meters. The Qushi'an River belongs to the plateau continental climate, with the average annual rainfall of 300-400 millimeters. The annual average runoff of the Qushi'an Damitan Hydrographic Station is 810 million cubic meters with an annual average surface water evaporation of 692.2 millimetres. The upstream of the river is canyon, and the middle and lower reaches are grassland. The basin is one of the main production place of the Qinghai province's Cordyceps sinensis.

The Daheba River is originated from the west of the Xinghai County, it runs from northwest to southeast into the Yellow River. The valley basin area is 4,028.89 square kilometres with an elevation of 3,000-4,700 meters. The Daheba River belongs to plateau continental climate, with an average annual rainfall of 300-400 millimeters. The annual average runoff of Daheba River Village hydrological station is 360 million cubic meters. On both sides of the Daheba River, there are desert basin due to lack of water.

The Ba Qu basin is originated from the west of the Zêkog (the Zeku) County, the east of the Tongde County. Two main tributaries are the Bagou and

## Hydrological Conditions and Water Resources

the Ga Qu, which flow from west to east into the Yellow River after convergence. The Ba Qu valley basin area is 4,286.78 square kilometers with an elevation of 2,800-4,200 meters. The Ba Qu belongs to plateau continental climate with an average annual rainfall of 400-500 millimeters. The annual average runoff of th Ba Qu is 320 million cubic meters, and the natural drop is 1,208 meters. The upstream flows through the plateau grassland, water resource there is inefficient, and the middle reach of the river is in a seasonal river. A hydropower station has been built in the basin, with an installed capacity of 1,600 kilowatts.

The Mang Qu is originated from the eastern part of the Guinan County, it runs from west to east into the Longyangxia Reservoir of the Yellow River with a basin area of 2,997.14 square kilometres and an elevation of 2,800-3,800 meters. Mang Qu belongs to plateau continental climate with an average annual rainfall of 300-400 millimetres. Its annual average runoff is 150 million cubic meters.and the natural drop is 1,524 meters. There are two hydropower stations in this basin, with a total installed capacity of 1,800 kilowatts. The basin is an important agricultural production base for the Guinan County.

The Longwu River is originated from the west of the Zêkog (the Zeku) County, it flows from south to north through the Tongren County into the Yellow River. The basin area is 5,035.40 square kilometres, and the terrain is high in the west and low in the east, with an altitude of 2800-4000 meters. The Longwu River basin belongs to plateau continental climate with an average annual rainfall of 300-500 millimetres. The annual average runoff of the Longwu River hydrological station is 610 million cubic meters with a natural drop of 2,292 meters. Farming and animal husbandry is the dominant industry in this drainage area, with wheat and barley as its main crops.

## II  Water System in the Yangtze River Source Area

The Ulan Moron (the Tuotuo River), the source of the Yangtze River is originated from the middle of the Geladaindong Peak of the Tanggula Mountains. It is called the Tongtian River after converge in the Nangjibalong with the Dang Qu in the south, and Chuma'er River in the north. It is also called the Tongtian River after flowing to the southeast into the Batang River in the Yushu County. The drainage area of the Yangtze River basin in the Sanjiangyuan is 158, 000 square kilometres, accounting for 40% of the total area of Sanjiangyuan. The main stream is 1,206 kilometers, accounting for 19.1% of the total length of the main stream. It has a drop of 2,065 meters, and the average drop is 1.78 ‰. In Sanjiangyuan, there are 109 First-Class tributaries of the Yangtze River above 50 square kilometers, of which there are 4 (the Dang Qu, the Chuma'er River, the Min River, and the the Yalong River) with the basin area of more than 10,000 square kilometers, 7 tributaries with an araa of 5,000-10,000 square kilometers, 22 tributaries with an araa of 1,000-5,000 square kilometers, and 28 tributaries with an araa of 500-1,000 square kilometers. The main tributaries of the Yangtze River source area are shown in Figure 2.11. Among the tributaries of the Yangtze River, those with large basin area and runoff include the Dang Qu basin, the Chuma'er River basin, the Zhamu Qu basin, the Mo Qu basin, the Beilu River basin, the Keqian Qu basin, the Sewu Qu basin, the Nieqia Qu basin, the De Qu basin, the upper reach of the Lancang River basin, and the upper reach of the Min River basin.

The Tongtian River, refers to the section for the upstream of the Yangtze River, especially to the Yangtze River section from converge area of the Ulan Moron (the Tuotuo River) and the Dang Qu to the Batang River entrance in the

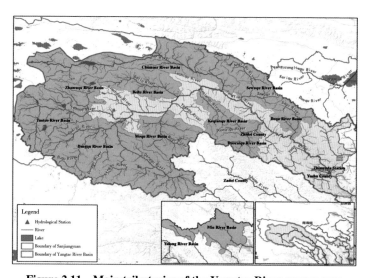

Figure 2.11 **Main tributaries of the Yangtze River source area**

## Hydrological Conditions and Water Resources

Yushu County. The Tongtian River is located in the Yushu-Tibetan Autonomous Prefecture which is in the southwest of Qinghai Province, flowing through four counties— Qumarlêb (the Qumalai County), Zhidoi (the Zhiduo County), Chindu (the Chengduo County) and the Yushu County. Located in the "roof of the world"— the Qinghai-Tibetan Plateau, the river has been descriped as passing to the sky, so it is called the Tongtian River. From the main source of the Ulan Moron (the Tuotuo River) and southern source of the Dang Qu, the Tongtian River flows to the east into the Zhiduo County, to Mo Qu, and then into the boundary river of the Zhiduo County and the Qumarlai County, and then not far from the Keqian Qu it turns to the southeast into the Chuma'er River in Deliechulabaden, at last becoming a surging river out of the Yangtze River source region. Continue to flow to the southeast, below the De Qu, it becomes border river of two counties—the Yushu County and the Chengduo County, and after flowing into the Batang River, it is called the Jinsha River. The main stream of the Tongtian River appears in a 弓 form (bow-shaped) with a total length of 813 kilometers. The river controls a basin area of nearly 140,000 square kilometers with an average annual flow of 400 cubic meters/second. The annual runoff is about 13 billion cubic meters with sediment volume of more than 9 million tons, the sediment content is 0.74 kilogram/cubic meter. The river water is clear with quality.

In June 2016, a comprehensive scientific expedition team of the Sanjiangyuan National Park made a hydrological survey to the Sanjiang source area. They tested a total of 12 corss-sections of the flow and measured hydrological elements in 10 rivers. The locations of the flow sections are shown in Figure 2.12.

The expedition team selected two sections in the Tongtian River for measurement. The Tongtian River No. 1 flow section is located in the Tongtian River Bridge along the National Highway 309 Line in the Zhiduo County. The geographical coordinates is Longitude 95°49'25" East, Latitude 34°02'12" North. The section of the river has a river bed of sand and gravel. In the upstream of the river, water is relatively concentrated, while in the downstream, water begins to diverge, and forms two streams. The left bank is steep, while the right bank is gentle with gravel river beach. The surrounding terrain is flat and open with good vegetation. Flow test is conducted by the ADCP. On the day of inspection (June 24, 2016), the water width of test

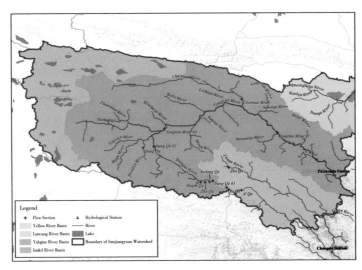

**Figure 2.12  The location of the measured flow cross-section of the Sanjiangyuan National Park**

section is of 224 meters, with an area of 225 square meters and flow of 256 cubic meters/second (See Figure 2.13).

The Tongtian River No. 2 flow section is located in the Nangjibalong Highway Bridge in the Zhiduo County. The geographical coordinate is Longitude 93°01'27" East, Latitude 34°08'56" North, with an altitude of 4,463 meters. The section of the river has a river bed of gravel, and the water is more concentrated. The upstream straight river is about 220 meters, while the downstream straight river is about 410 meters. The height between bridge and the water surface is 7 meters, the two sides' slopes are gentle with sand and gravel, and the surrounding terrain is flat. The vegetation is good. The flow test is conducted by the ADCP. On the day of inspection (June 25, 2016), the test section is 165 meters wide with a cross-sectional area of 213 square meters and the discharge is of 260 cubic meters/second (See Figure 2.14).

The Dang Qu, also known as the Akedamu River, is the southern source of the Yangtze River, which is located in the southwest corner of Qinghai Province. It is originated from the north branch of the Xiasheriaba Mountain in the eastern section of the Tanggula Mountains in the Zaduo County. It flows

# Hydrological Conditions and Water Resources

(a) Picture of flow section of the Tongtian River Bridge

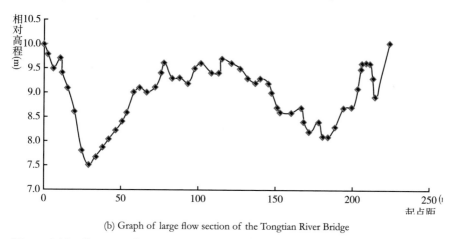

(b) Graph of large flow section of the Tongtian River Bridge

**Figure 2.13  Cross-section profile of the Tongtian River Bridge Hydrological Station**

through the Yushu-Zadoi County and the Tanggula Township of the Geermu City, making convergence in the source of the Tuotuo River in Nangjibalong in the Zhiduo County. Below the convergent area, it is called the Tongtian River. The length of the Dang Qu is 352 kilometres, it has a basin area of 31,269.43 square kilometres and an elevation of 4,600-5,100 meters. The average annual flow is 146 cubic meters/second, and the annual flow is 4.602 billion cubic

(a) Picture of flow section of Nangjibalong of the Tongtian River

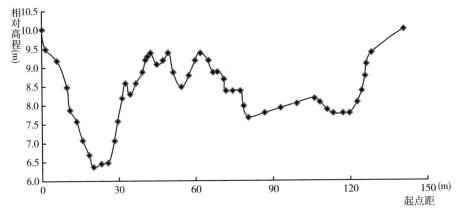

(b) Graph of large flow section of Nangjibalong of the Tongtian River

**Figure 2.14　Cross-section of the Nangjibalong in the Tongtian River Hydrological Station**

meters. The Dang Qu belongs to continental monsoon climate. Affected by the geographical environment, the annual climate is cold with no four seasons and the average annual rainfall is 700-900 millimetres. Ethnic groups in the drainage area are mainly Tibetan, as well as some Han, Hui, Mongolian and so on. The basin is a natural pasture with rich aquatic plants, and the people working in the first industry account for more than 90%. Between agriculture and animal husbandry, the local government gives priority to animal husbandry.

The comprehensive scientific expedition team of the Sanjiangyuan National Park selected two sections for measurement in the Dang Qu. The Dang Qu No. 1 flow section is located at 105 meters above the upper reaches of the Highway Bridge in Zadoi County, the geographical coordinate is of Longitude 94°12'54" East, Latitude 32°52'11" North and its altitude is 4,740 meters. The surrounding mountain grassland is flat with good vegetation. The section of the river bed is sand and gravel. It flows steadily and concentratedly. The left bank is grassland, while the right bank is gravel sand beach. Flow test is conducted by rod suspended current meter. On the inspection day(June 22, 2016), the surface width of test section is 8.9 meters, with a cross-sectional area of 2.15 square meters, and a maximum water depth of 0.36 meters. The maximum discharge is 0.35 meters/second, with a flow of 0.581 cubic meters/second (See Figure 2.15).

(a) Picture of upstream of the Dang Qu at the Zhadan Highway Bridge and flow section

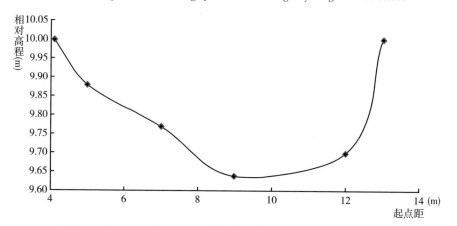

(b) Graph of upstream flow section of the Dang Qu at the Zhadan Highway Bridge

**Figure 2.15  Upstream cross–section profile in the Zhadan Highway Bridge of the Dang Qu Hydrological Station**

The Dang Qu No. 2 flow section is located in the upstream highway bridge, 30 kilometres from the Gari Qu in the Suojia rural township, the Zhiduo County. The geographical coordinate is of Longitude 92°45'34" East, and Latitude 33°42'40" North. It is located in Yanshiping, which is 75 kilometers away from the National Highway 109 Line. The section is located at 4,510 meters, and the catchment area is 16,100 square meters, shich is 61 kilometres to Nangjibalong. The river bed is basically sand and gravel. The river flow is traight, steady and concentrated. The flow test is done by the ADCP. On the day of inspection (Junc 26, 2016), the test section is 80 meters wide with a cross-sectional area of 86.4 square meters, the average flow rate is 1.28 meters/second and the discharge is 111 cubic meters/second (See Figure 2.16).

(a) Picture of the Dangqu River Bridge and flow section

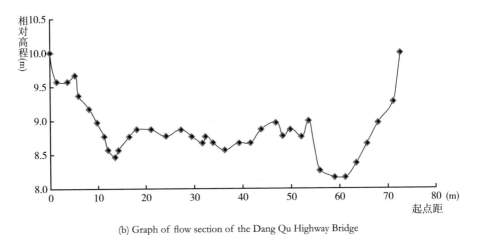

(b) Graph of flow section of the Dang Qu Highway Bridge

**Figure 2.16　Cross−section profile of the Dang Qu Highway Bridge Hydrological Station**

## Hydrological Conditions and Water Resources

The Chuma'er River, the north source of the Yangtze River, is originated from the east of the Hoh Xil (the Kekexili) Mountain, which is in the southern part of the Kunlun Mountains, and it is in the west of the Zhiduo County. There are two sources: the north branch is originated from the place about 18 kilometers southeast of the Kekexili Lake. Its source locatess 4,920 meters high with a total length of 46 kilometers. The west branch is originated from the south of the Heiji Mountain, which is in the southern part of the Kekexili Lake with an altitude of 5,432 meters, and its total length is 45.7 kilometers. After the convergence of the two streams, it flows 94 kilometers to the east into the Yelusu Lake (also known as Duoergaicuo or Cuorendejia), and continues to flow 20 kilometers through the lake, and then flows through 117 kilometers to the Chuma'er River along the Qinghai-Tibetan Highway Bridge into the Qumalai County. The average annual flow is 7.78 cubic meters/second. The river continues to flow eastward, and the downstream part turns gradually to the south into the Qumatan. To the south of the Quma River Township, the river flows into the Tongtian River at the Laiyong Beach. The estuary elevation is 4,216 meters. The length of the Chuma'er River is 515 kilometers, and the basin area is 20,800 square kilometers. The average annual flow is about 33.0 cubic meters/second with an annual runoff of 928 million cubic meters. The Chuma'er River Basin terrain is high in the west and low in the east with an altitude of 4600-5200 meters, it belongs to plateau continental climate. The average annual rainfall is 200-300 millimetres. The upper reaches of the basin is a part of the Kekexili, which is famous for its non-inhabited area. The place possesses various national protected animals--Tibetan antelope, Tibetan wild donkeys, cattle and other animals.

The comprehensive scientific expedition team of the Sanjiangyuan National Park selected a section for measurement in the Chumar'er River Bridge. The geographical coordinates is Longitude 94°56'32" East, and Latitude 34°51'19" North.

The left bank of the section is alluvial, the right bank is close to the mountain. The water flows along the mountain, and the river nearby is sparsely distributed like a mesh. There are three bridges built respectively on the upper, middle and lower part of the river. The section elevation is 4,270 meters. Under the first bridge, water is relatively concentrated. When the water level is low, the left bank is dry. The main flow concentrated in the right bank with a water surface width of 70-90 meters, and the distance from the mainstream to the right bank of the bridge is about 26-50 meters. The flow test was conducted via using the ADCP. On the day of inspection (June 24, 2016), the test section is 41 meters wide with a cross-sectional area of 9.8 square meters. The discharge is 5.42 cubic meters/second (See Figure 2.17).

The Zhamu Qu, originated from the Tanggulashan Town and the border of the Zhiduo County, flows from north to south into the Yangtze River

(b) Graph of river flow section the Chuma'er River and the Qu Ma Bridge

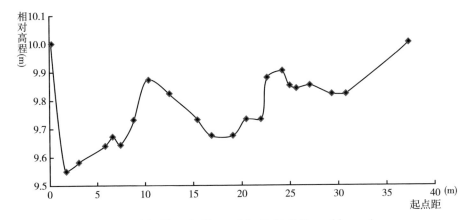

(a) Picture of the Chuma'er River and the Qu Ma Bridge and fow section

**Figure 2.17　Cross–section profile of the Chuma'er River and the Qu Ma Bridge**

## Hydrological Conditions and Water Resources

(the Tongtian River). The valley basin area is 5,293.98 square kilometers with an elevation of 4,600-5000 meters. The Zhamu Qu valley basin belongs to temperate semi-humid plateau monsoon climate, which is cold in winter and cool in summer, with abundant rainfall and relatively less sunshine. The average annual rainfall is 800-900 millimeters. The annual average flow of the Zhamu Qu is 240 million cubic meters, and the natural drop is about 647 meters.

The Mo Qu drainage area, adjacent to the Dang Qu basin, is originated from the northern part of the Narigen Mountains in the Zaduo County. It runs from south to north into the Yangtze River (the Tongtian River) through the Zhiduo County. The drainage area is 8,892.04 square meters, and the terrain is high in the south and low in the north with an altitude of 4,500-5000 meters high. The Mo Qu basin belongs to plateau continental climate with an average annual rainfall of 700-800 millimeters. The annual average runoff of the Mo Qu is 690 million cubic meters, and the natural drop is 610 meters.

The Beilu River is originated from the southwest of Gongchangrima in the west of the Qumalai County, flows from west to east into the Yangtze River (the Tongtian River). Within the drainage basin, there are many rivers. The river basin area is 7,962.82 square kilometers and the altitude is 4300-5000 meters.

The Beilu River basin belongs to plateau alpine climate, and the average annual rainfall is 800-900 millimeters. The natural drop of the river is 613 meters, and the average annual runoff is 660 million cubic meters. The upper part of the river is called the Riachi Qu, which is a seasonal river. In the lower reach, perennial water exists. Most parts of the river passes sandy beach, and sparsely populated regions.

The Keqian Qu is originated from the middle of the Zhiduo County, and runs from south to north into the Yangtze River (the Tongtian River). The basin area is 3,567.72 square kilometers with an elevation of 4300-5200 meters. The Keqian Qu belongs to plateau continental climate and the average annual rainfall is 600-800 millimeters. The natural drop of the river is 1,085 meters, and the average annual runoff is 280 million cubic meters. The ethnic groups in this basin are mainly Tibetan, there are also Han, Hui, Mongolian people and so on. Animal husbandry is the dominated economy in this region.

The Sewu Qu is originated from the middle of the Qumalai County, and is adjacent to the source of the Yellow River in the north. There are two main tributaries of the Angri Qu and the Sewu Qu, making confluence from northeast to southwest into the Yangtze River. The drainage area is 6,753.69 square kilometers with an elevation of 4500-8000 meters. The Sewu Qu belongs to plateau alpine climate, and the average annual rainfall is 700-800 millimeters. The natural drop of the Sewu Qu is 536 meters, and the average annual runoff is 540 million cubic meters. There are many mountains and valleys in this basin, with good vegetation coverage and developed animal husbandry.

The Nieqia Qu is originated from the Ajiajima Peak in the junction of Zaduo County and the Zhidoi County (5930 meters in elevation), and it flows from southwest to northeast into the Yangtze River. The drainage area is 5,695.29 square kilometers, with an altitude of 4400-6000 meters. The Nieqia Qu basin belongs to plateau continental climate with an average annual rainfall of 550-650 millimeters. The Nieqia Qu flows through the Zhiduo County. The residents are mainly Tibetan, as well as some Han, Hui, and Mongolian. At the downstream of the Nieqia Qu, there is a water diversion power station built.

The De Qu is originated from the border of the Qumalai County and the Chengduo County, and it runs from north to south into the Yangtze River. The drainage area is 4,234.60 square kilometers with an elevation of 4300-5000 meters. The De Qu basin belongs to cold temperate continental monsoon climate with an annual rainfall of 550-650 millimeters. The natural drop of the De Qu is 924 meters, and the average annual runoff is 940 million cubic meters. The land is rich in swamps with intermittent flow. The drainage area is well covered by vegetation and animal husbandry is well developed.

The main stream of the Yalong River and its tributary the Xianshui River are both originated from the Sanjiangyuan area. The main stream is originated

## Hydrological Conditions and Water Resources

from the south of the Bayan Har (the Bayankela) Mountains in the Chengduo County. The Xianshui Rriver is originated from the southern part of the Bayan Har (the Bayankela) Mountains in the Dari County. The total area is about 12,000 square kilometers, with an elevation of 4200-4900 meters.

The Min iver, a tributary in the middle reach of the Yangtze River, is originated from the Sanjiangyuan area. The Min River Basin in the Sanjiangyuan area is located in the Dari County, the Jiuzhi County, and the Banma County, with an area of about 9,775 square kilometers and an elevation of 3600-4400 meters. There are many river networks, including the Dadu River, the Duke River, the Ma'er Qu, the Ake River and some other rivers in this basin.

### Ⅲ Water System in the Lancang River Source Area

The Lancang River is originated from the south of Chajiarima of the northern part of the Tanggula Mountains. The source of river has an elevation of 5,388 meters, and the main stream in Qinghai is called the Zha Qu. The Lancang River flows from the village of the Nangqian County into Tibet. The drainage area is 37,000 square kilometers, accounting for 9.4% of the total area of

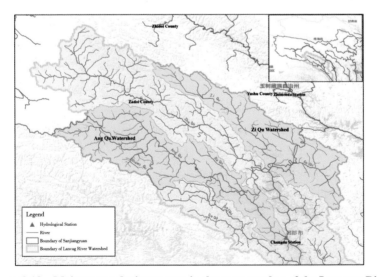

Figure 2.18 Mainstream drainage area in the uper reaches of the Lancang River

125

Sanjiangyuan, with rivers of 448 kilometers long, accounting for 20.4% of the total length of the main stream and the total drop is 1,553 meters, with the average drop of 3.47 ‰. The water networks is well-developed and dendritic there. In addition to the mainstream--the Zha Qu, the Ang Qu and the Zi Qu are also originated from Sanjiangyuan.

The Zha Qu, also called the Za Qu, is originated from high ground which is 4 kilometer away from the west of Chajiarima in the north of the Tanggula Mountains which is located in the northwest of the Zaduo County in the Yushu-Tibetan Autonomous Prefecture. The altitude of the river source is 5,388 meters. The main stream which is from the river source to the mouth of the Longmao Qu is called Jiaguokongsanggongma Qu. Up until Ganasongduo, it is called the Zana Qu, and from here downwards it is called the Zha Qu. The river flows from northwest to southeast through the Nangqian County into Tibet Autonomous Region. In Qinghai Province, it has 448 kilometres-long main stream, with a total length of 4.5 kilometres long river shared by both Qinghai and Tibet. The drainage area is 18,500 square kilometres. The average annual flow is 138 cubic meters/second, and the average annual runoff is 4.352 billion cubic meters.

The Comprehensive Scientific Expedition team of Sanjiangyuan National Park selected the Zhaqu Bridge River section for the survey. The geographical coordinate is of Longitude 94°36'44" East, Latitude 33 °11'58" North, and an altitude of 4,370 meters high. The section of the river bed is sand and gravel. The water under the bridge diverges, and it converges 60 meters downstream of the bridge. The right bank is mainly gravel sand and gravel, and the left bank is grass. The surrounding terrain is flat and open with good vegetation. Flow test is conducted by the ADCP. On the day of inspection (June 2016), water surface width of the tested section is 47 meters with a cross-sectional area of 42.7 square kilometers, and the discharge is 61.2 cubic meters/second (See Figure 2.19).

The Ang Qu is originated from the northern part of the Baqing County in Tibet, and the western part of the Zaduo County in Qinghai. After running across the Zaduo County, the Nangqian County and the Leiwuqi County, the

# Hydrological Conditions and Water Resources

(a) Picture of the Zha Qu and Zhaaqu Bridge's cross-section

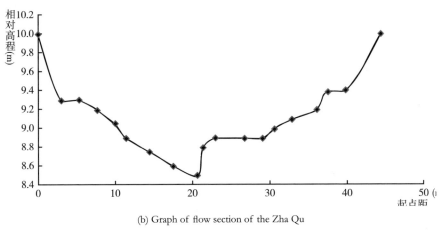

(b) Graph of flow section of the Zha Qu

**Figure 2.19   Cross−section profile of the Zha Qu Bridge Hydrological Station**

river flows into the Lancang River in the Changdu County. The dainage area is 16,800 square kilometres, and the terrain is high in the west and low in the east, and the elevation is 3600-5200 meters high. The average flow of the Ang Qu for many years is 186 cubic meters/second, with a natural drop of 1,898 meters. The Changdu County is an economic center in the eastern Tibet, because of its convenient transportation, frequent exchanges of material and cultural.

The Zi Qu, originated from the southeastern hills of the Zaduo County, runs through the Nangqian County, into the Lancang River in Tibet. The basin area is 12,900 square kilometers. The annual average flow of the Zi Qu is 137 cubic meters/second, and the natural drop is 1,540 meters. The Zi Qu Basin drainage area develops well, with many river networks and rich water resources.

## IV  Water System in the Inland Drainage Area

Located in the northwest of the Sanjiangyuan area, the Qiangtang Plateau is China's famous inner lake area where river networks are well developed with many lakes. Due to the terrain barrier, although there are no direct pathways between the rivers and lakes in this region and the external rivers-the Yangtze River and the Lancang River, these rivers and lakes are still closely linked because of the exchange of the groundwater, atmospheric water and heat exchange. Figure 2.20 shows the river and lakes in the Qiangtang Plateau.

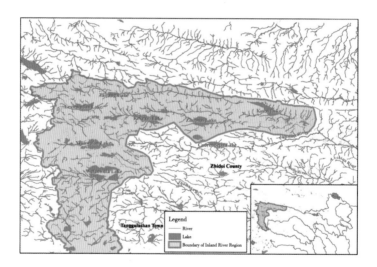

Figure 2.20  Rivers and lakes in the inland lake area of the Qiangtang Plateau

## V  River Runoff

The rivers in the Sanjiangyuan area are divided into two major categories: the outflow river and the inner stream river. The average annual runoff of the three rivers is 52.29 billion cubic meters, of which water flowing out of the Yellow River is 21.06 billion cubic meters, water flowing out of the Yangtze

Hydrological Conditions and Water Resources

River is 18.63 billion cubic meters, and water flowing out of the Lancang River is 12.6 billion cubic meters. Rainwater and snow melt water is the main source of therunoff.

The Tangnaihai hydrological station is an important controlling station in the upper reach of the Yellow River, regulating 121,972 square kilometers of basin area, and 1,553 kilometers of river above this section, accounting for 28.4% of the whole river length. It is the storage station of Longyangxia. From 1956 to 2015, the multi-annual mean runoff of the Tangnaihai station is 20.1 billion cubic meters, accounting for 37.7% of the Yellow River's annual average natural runoff. The maximum annual runoff is 32.84 billion cubic meters in 1989, and the smallest is 10.64 billion cubic meters in 2002, a change of 22.2 billion cubic meters lies in between. The annual runoff decreased after 1983, and water began to rise after 2003. The distribution of the Yellow River runoff within a year is very uneven, the runoff of the flood season (between May and October) accounts for about 78% of the total annual runoff. The river runoff in the source region of the Yellow River is mainly from precipitation, ice, snow melt water and groundwater recharge also contribute.

Since the 1980s, the amount of water from the Yellow River to the sea has significantly reduced. In fact, the amount of water in the Yellow River source does not keep decreasing. As shown in Figure 2.21, the amount of water in the Tangnaihai Station in the source area of the Yellow River fluctuates year by year, but it has remained basically at a level of 20 billion cubic meters in the long run, but the runoff of the Huayuankou Hhydrological Station in the Yellow River estuary has obviously decreased. The life of the Yellow River is at the its source, so the ecological significance of the Sanjiangyuan region is enormous. If not properly protected, our mother river— the Yellow River will suffer serious ecological disasters.

The Longwu River estuary is the outlet of main stream of the Yellow River when it flows out of the Sanjiangyuan area. Hydrological station here has been removed. Based on the statistics of runoff from the nearest controlling station (Guide (No.2) station), the amount of outflow water of the Yellow River from the Sanjiangyuan area is measured by enlarging the data according to the drainage area. By calculation, the average annual outflow runoff of the Yellow River mainstream from the Sanjiangyuan area is about 21.06 billion cubic meters.

The Zhimenda Hydrological Station is an important control station in the upper reach of the Yangtze River, and it holds a basin area of 137,704 square kilometers. The average annual flow of the Zhimenda hydrological station during 1956-2015 is 12.98 billion cubic meters, accounting for 1.3% of the average annual runoff of the Yangtze River. The maximum is 24.6 billion cubic meters in 2009, and the minimum annual runoff is 7.04 billion cubic meters in 1979, a change of 17.56 billion cubic meter lies in between (see Figure 2.21). The annual runoff is basically stable before 1994, and the trend of runoff growth after 1994 is more significant. The runoff distribution of the Yangtze River during these years is very uneven, with the flood season (from May to October) runoff accounting for about 87% of the total annual runoff. In addition to natural precipitation, the main source of the runoff in the Yangtze River source area is glacier melting water. The interannual variation of the runoff is mainly affected by precipitation and climate, and the change within a year is mainly affected by the temperature and precipitation process, there are more rainfall during summer and autumn seasons.

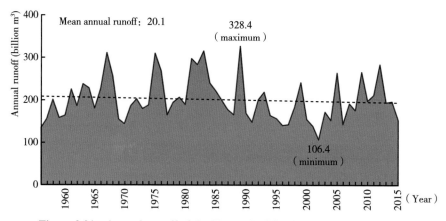

Figure 2.21　Annual runoff of the Tangnaihai Station in the Yellow River

# Hydrological Conditions and Water Resources

The Zhimenda Hydrological Station is located closely to the outlet of mainstream of the Yangtze River in the Sanjiang source area. After enlarging in proportion, the average annual outflow of water is 14.0 billion cubic meters in the station. In addition, the Yalong River and the Min River, two tributaries of the Yangtze River, also contributed 1.52 billion cubic meters and 3.31 billion cubic meters of outflow, respectively. Therefore, the average annual runoff of the Yangtze River outflow from the Sanjiangyuan area is 18.63 billion cubic meters.

According to the survey results of Qinghai Province water resources in 2006, the average annual runoff of the Lancang River in Sanjiangyuan area is 12.6 billion cubic meters, accounting for about 15% of the total runoff. Surface runoff is relatively stable with rich water resources.

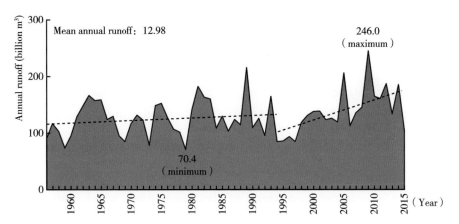

**Figure 2.22　Annual runoff of the Zhimenda Station in the Yangtze River**

## Section 3　Wetland Resources

The Sanjiangyuan area is rich in wetland resources. According to the results of the 2$^{nd}$ Wetland Resources Survey in Qinghai Province released on June 6, 2014 by Qinghai Provincial People's Government Information Office and the Qinghai Provincial Forestry Bureau, the total area of wetland in the Sanjiang Park is 41,700 square kilometers. Among them, the Yangtze River wetland area is 19,000 square kilometers, accounting for 45.6%; the Yellow River wetland area is 12,200 square kilometers, accounting for 29.3%; the Lancang River wetland area is 1,400 square kilometers, accounting for 3.4%; other influx of wetland area is 9,100 square kilometers, accounting for 21.8%. According to the origin and characteristics of wetland, it can be divided into four categories: lake wetland, marsh wetland, river wetland and artificial wetland. The percentage of the four wetlands in the Sanjiangyuan is 21.1%, 63.7%, 14.2% and 1.1%, respectively in area.

Lakes wetlands are wetlands consisting of natural low-lying land with different shapes and sizes of water on the ground. Qinghai Province is a place with lots of lakes, mainly distributed in the Qinghai Lake, Kekexili, the upper reaches of the Yangtze River, the Yellow River and the lake areas along the southern Qaidam (Chaidamu) Basin. The total area of lake wetland is 8,775 square kilometers, of which 1,393 square kilometers is in the Yangtze River Basin, 1,551 square kilometers is in the Yellow River Basin, 3,582 square meters in the Qiantang Plateau lake groups area. The wetlands included in the list of important wetlands in China are the Gyaring (the Zhaling) Lake, the Ngoring (the Eling) Lake, the Madoi (the Maduo) Lake, the Gangnagemacuo Lake in the Yellow River source area, the Yirancuo Lake in the Yellow River source area, and the Duoergaicuo Lake in the Yellow River source area. Among them, the Zhaling Lake, and the Eling Lake are the two largest freshwater lakes in the main stream of the Yellow River, which have a huge potential in water regulation.

The Eling Lake, a rifted tectonic lake, is the largest freshwater lake in the Yellow River Basin. The lake is located in te Maduo County, Qinghai Province, covering an area of 650.75 square kilometers. With about 34 kilometers wide from east to west and 34.5 kilometers long from north to south, it appears in an upstanding triangle form. Water storage capacity of it is about 11.6 billion cubic meters, with deep water in the north and shallow water in the south. The average depth is 17.81 meters, with the maximum depth of 30.7 meters in the north of the central part of the lake. The bottom of the Eling lake is vast and flat. The Lake is clear with blue colour and high water quality of and a pH value of 8.05-8.40.

The Zhaling Lake is the second largest freshwater lake in the Yellow River Basin. It is a tectonic lake formed by rift basins, separated from the east of the Eling Lake in more than 10 kilometers by a mountain. The Zhaling Lake covers an area of 544.73 square kilometers, about 37 kilometers from the east to the west and 23 kilometers from the north to the south. It appears in an upside down triangle form. Water storage capacity of the lake is about 4.9 billion cubic meters, with deep water in the north and shallow water in the south. The average depth is 9 meters; the maximum depth is 13.1 meters in the northeast of the central part of the lake. The lake is clear with gray colour and high water quality, the pH value is 8.35-8.50.

Table 2.2 shows the surface area of the main lakes in the Sanjiangyuan area. The data are derived from the global surface cover data (GlobeLand30) published by the National Basic Geographic Information Center in 2017. The maximum value of each lake area in 2010 is counted.

Table 2.2  Main lakes surface statistics of Sanjiangyuan (based on GlobeLand 30)

| Lake | Area (square kilometers) | Lake | Area (square kilometers) |
| --- | --- | --- | --- |
| The Qinghai Lake | 4362.96 | The Kekexili Lake | 320.97 |
| The Eling Lake | 685.15 | The Kusai Lake | 284.06 |
| The Wulanwula Lake | 591.98 | The Zhuonai Lake | 270.92 |
| The Chibuzhangcuo Lake | 550.35 | The Cuorendejia Lake | 249.64 |
| The Zhaling Lake | 530.42 | The Donggeicuona Lake | 234.25 |
| The Xijin Wulan Lake | 419.29 | The Duoergai Lake | 228.86 |
| The Longyangxia Reservoir | 375.8 | The Taiyang Lake | 102.23 |

At the same time, the Qinghai Provincial Department of Water Resources has conducted a field survey on the Qinghai Lake, the Eling Lake and other major lakes. The main data is shown in Table 2.3. It shows there is no big difference between the data provided by GlobeLand30 and the data got from field survey. So the accuracy of data provided by GlobeLand30 is high.

Table 2.3   The hydrological characteristic of main lakes in Qinghai Province based on field survey

| Lake | Average water Level (meters) | Average water depth (meters) | Maximum depth of water (meters) | Area (square kilometers) | Volume (hundred million cubic meters) |
| --- | --- | --- | --- | --- | --- |
| The Qinghai Lake | 3193.50 | 18.30 | 26.60 | 4294.00 | 785.20 |
| The Eling Lake | 4270.10 | 17.81 | 32.09 | 650.75 | 115.91 |
| The Zhaling Lake | 4290.80 | 9.00 | 13.54 | 544.73 | 49.03 |
| The Donggei Cuona Lake | 4084.90 | 28.78 | 90.22 | 247.13 | 71.12 |

Note: The Zhaling Lake includes the Chamucuo Lake and the Zhuorangcuo Lake.

In the past 30 years, the number and area of lakes in both of the southeastern part of the Yellow River and the Yangtze River are shrinking in large scale, and some lakes even disappeared in the northwestern part of the Yellow River, with lakes transforming into grassland. Some big lakes are split into several smaller lakes. On the contrary, the western and northern lakes in the Yangtze River source area are in expansion, with some low-lying areas even forming into some new lakes. From the late 1970s to the early 1990s, except for the southwestern part of the Yangtze River source area, the main rivers in the Sanjiangyuan area shows the trend of shrinking, especially in the western part of the source area. From the early 1990s to the beginning of 2004, the early shrinking lakes began to expand, especially in the northwestern part of the Yangtze River source area. But the lakes in the Yellow River source area have been in a continuous shrinking trend.

Marsh plays an important role in humidifying climate, regulating river runoff, replenishing groundwater and maintaining regional water balance. The environment in the Sanjiangyuan area is harsh with many special types of swamp. The main types of swamps are the three-leafed cauliflower swamp and the fir leaf almonds, of which most belongs to peat marsh. In the Yellow River source, and the three Yangtze Source Rivers—the Tuotuo River, the Chuma'er River and the Dang Qu River, as well as the Lancang River source, there are large swamps, with a total area of 26,500 square kilometers.

The Yangtze River basin has a swamp area of 14,000 square kilometers. While the swamps mostly concentrat in the humid eastern and southern part

Hydrological Conditions and Water Resources

of the Yangtze River source area, but a rather small area in the dry western and northern part. From a terrain point of view, the swamps are mainly distributed in the low-lying areas of the riverside lakes, especially in the middle and upper reach of the river. The swamps are mainly in the middle and upper reach of the Dang Qu, and the upper reaches of the Tongtian River. On the northern side of the Tanggula Mountains, the highest swamp has grown to an altitude of 5,350 meters, reaching the upper limit of the Qinghai-Tibetan Plateau, which is the highest swamp in the world now.

The Yellow River Basin has a swamp area of 9,037 square kilometers, and the marsh is affected by semi-arid conditions. It is mainly distributed in the Yueguliezong Qu, surrounding two lakes and the Xinxiuhai area. The total area of the swamp in the source area of the Lancang River is 986 square kilometers, it mainly concentrats in the upper reach of the main stream of the Zha Qu and the tributary of the A Qu. There is a swamp area of about 2500 square kilometers in other inland drainage areas.

In addition to lake wetlands and marshes wetlands, the river wetland area in the Sanjiangyuan area covers 5,898 square kilometers, and artificial wetland covers 444 square kilometers.

## Section 4  Glacier Snow Mountain

Glacier is a unique mountain landscape in western China where glacier melting water supplies river, and irrigates inland basin and farmland. The process of glacier melting regulates the climate, and reduces the local temperature. The Sanjiangyuan area has 715 glaciers in total, about 2,400 square kilometers in the area, and 200 billion cubic meters in glacier resources reserves. The spatial distribution of the glaciers is shown in Figure 2.23. With most glaciers in the source of the Yangtze River, there are 627 glaciers, covering an area of 1,247.21 square kilometers, with 98.3 billion cubic meters of glacier deposits, and about 989 million cubic meters of ablation. In the source area of the Yellow River, there are 68 glaciers, covering an area of 131.44 square kilometers, glacier reserves up to 1.104 billion cubic meters; In the Lancang River source area,

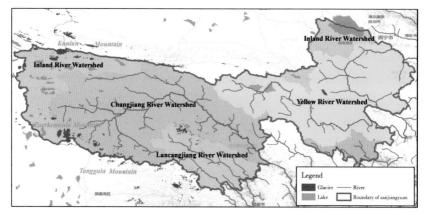

Figure 2.23   Distribution of glaciers in the Sanjiangyuan region

Figure 2.24   Glacier snow mountain view (the left is the Kunlun Mountain, the right is the Tanggula Mountain)

the number and basin area of glaciers are smaller, with only some 20, and an area of 124.12 square kilometers.

Modern glaciers in the Yangtze River source area belong to continental mountain glaciers. The glaciers are mainly distributed in the northern slope of the Tanggula Mountain, the western section of the Zu'erkenwula. The Kunlun Mountains with the Dang Qu basin covering the largest area of glaciers, are followed by the Tuotuo River Basin, and Chuma'er River Basin as the smallest. The largest snow-capped mountains glacier lies in Geladaindong, and Gaqiadirugang of the Tanggula Mountains and Gangqin of the Zu'erkenwula Mountains, where Geladandong is the most magnificent.

Hydrological Conditions and Water Resources

Table 2.4  Glacier area statistics in the Yangtze River Area

| Water System | Glacier Area (square kilometers) | | |
|---|---|---|---|
| | The Tanggula Mountain | The Kunlun Mountain | total |
| The Tuotuo River | 380.97 | | 380.97 |
| The Dang Qu River | 793.4 | | 793.4 |
| The Chuma'er River | | 54.99 | 54.99 |
| Upper Section of the Tongtian River | 12.40 | 5.45 | 17.85 |
| Total | 1186.77 | 60.44 | 1247.21 |

The Tuoluogang (5,014 meters above sea level), locates in the source of the Tuoluo Qu, a tributary of the Duo Qu in the middle of the Bayankela Mountain of The Yellow River Basin. There exists about 4 square kilometers residual glaciers in this region, and 80 million cubic meters glacier reserves. There are 14 mountains with an elevation above 5,000 meters, such as the Kali'enkazhuoma, the Manite, the Riji and the Lenadongze and etc. There reserves 140 million cubic meters mutil-year solid water. All together, in Tuoluogang there reserves 220 million cubic meters of water approximately. The annual melting water is about 3.2 million cubic meters, supplying rivers with melting water.

In the north of the Lancang River source area, there are many year-round snow peaks, with an average elevation of 5,700 meters. The highest peak is up to 5,876 meters. Among the snow peaks, there are the Quaternary mountain glaciers, which extend to 34 kilometers long from east to west, 12 kilometers wide from north to south. There are more than 20 glaciers with an area of more than 1 square kilometer. Affected by global warming and other factors, the glacier area in the Sanjiangyuan area has reduced 233 square kilometers as a whole in the past 30 years. The data shows that the retreat rate of glaciers in the

Dang Qu soure area reaches 9 meters per year, the retreat rate is 8.25 meters per year in the source of the Tuotuo River, and Gangjiaquba glaciers in Geladandong reduce 500 meters in the last 20 years, with a retreat rate of 25 meters per year. Below the Lancang River source line to the lowest permafrost zone, at the elevation of 4,500 meters to 5,000 meters, the land appears in periglacial landform. Due to increased heat at the lower part, ice-meilting induced slides, hot ice-melting induced depression and other types of lands appear. The north mountain slope is more than 2 times longer than that in south, with the ice tongue extending from the elevation of 5,800 meters down to the end of the elevation of 5,000 meters or so, and the longest ice-tongue is 4.3 kilometers. The largest glacier in the source area is the Sederi Glacier, covering an area of 17.05 square kilometers, which is supplied by water from the two small tributaries of the Qiongrinong and the Charinong of the Chari Qu.

## Section 5  Groundwater Resources

The Sanjiangyuan groundwater resources are rich in reserves. According to the findings of the 2006 Qinghai Province Water Resources and Resources Survey published in Qinghai Province water resources evaluation report, the groundwater resources in Sanjiangyuan are 19.33 billion cubic meters in total.

The amount of groundwater resources in the Yangtze River source area is about 7.12 billion cubic meters. The water belongs to hilly area groundwater, mainly the bedrock fractured water, followed by the leaking water from the loose rocks, and the frozen layer water. The recharge source is mainly supplied by the vertical precipitation of natural raining and the horizontal runoff water is supplied by glacier melting. The distribution of groundwater and precipitation is in consistence. In the source area of the Yangtze River, water out of the groundwater is widely distributed. Near the valley of the river tributaries, there are many dense spring networks widely distributed, which is especially abundant in the lower reach of the north of the Chuma'er River. Groundwater along the ruptured channel emerges out and forms hot springs, which are often exposed in the north and the south of the Yangtze River source area. There are many in the north of the Tanggula Mountains, and the most concentrated ones are in the upper section of the valley area of the Bu Qu. Groundwater in hilly area emerges out into surface water, so the amount of surface water resources in the Yangtze River source area equal alomost to the total amount of water resources.

The amount of groundwater in the source area of the Yellow River is about 6.61 billion cubic meters. The Yellow River source area belongs to plateau hilly area. The groundwater reserves as the fissured water in surface active layer of permafrost in mountains, it supplies the runoff and converts into surface water.

The amount of groundwater in the Lancang River source area is about 4.58 billion cubic meters. The Lancang River area belongs to the plateau hilly area, and its distribution is mainly rock-based fissure water, and some rock leakage water. The river is supplied by a single source, mainly the vertical precipitation and ice melting water. The horizontal runoff comes out through the river and undercurrent. The water has high quality with a pH value of 7-8.5.

In addition, the amount of groundwater reserves in inner stream of the Sanjiangyuan area is about 1.02 billion cubic meters, of which 300 million cubic meters is in the inner stream area of the Qiangtang Plateau, 190 million cubic meters is in the (Chaidamu) Qaidam Basin and 530 million cubic meters is in the Qinghai Lake drainage area.

# G.3
# The Course and Achievements of the Sanjiangyuan Ecological Environment Protection

As the Central Party and the State Council attach great importance to the ecological environment protection and construction of Sanjiangyuan, governments at all levels and the whole community have taken active actions with comprehensive measures including policy, project, capital and publicity and others. They make great efforts to protect the ecological environment of Sanjiangyuan. Through over ten years' ecological protection and governance, particularly powerful initiatives of the Qinghai Party Committee and the Qinghai Government, great achievements have been made in ecological environment protection and construction of Sanjiangyuan. Immense contributions have been made to guarantee China's ecological safety and promote ecological civilization construction.

## Section 1  Establishment of the Sanjiangyuan Comprehensive Pilot Zone and National Park

The "Sanjiangyuan Nature Reserve" is an important initiative adopted by the State Forestry Administration and Qinghai Province in order to proactively respond to the great call for "rebuilding the beautiful landscape of the great northwest" and carry out the western development strategy proposed by the Central Party and the State Council. The Qinghai People's Government approved the establishment of the Sanjiangyuan Provincial Nature Reserve in May 2000 with a total protection area of 152,300 square kilometers, accounting for 21% of the total area of the province and 42% of the

# The Course and Achievements of the Sanjiangyuan Ecological Environment Protection

Sanjiangyuan total area, and involving six counties in the Guoluo-Tibetan Autonomous Prefecture, six counties in the Yushu-Tibetan Autonomous Prefecture, two counties in the Hainan-Tibetan Autonomous Prefecture, and two counties in Huangnan-Tibetan Autonomous Prefecture. The Tanggula Mountain Town of the Geermu City, and 16 counties and one town in all are also involved. It consists of 69 incomplete villages and towns in terms of administrative division.

In order to build the Sanjiangyuan Nature Reserve into an international natural reserve, the Qinghai People's Government decides to apply for the Sanjiangyuan National Nature Reserve and organizes the compilation of General Plan for the Sanjiangyuan Nature Reserve in 2001. In January 2003, the State Council (General Office of the State Council [2003] No. 5 document) approved the Sanjiangyuan National Nature Reserve (hereinafter referred to as the Reserve). Being part of Sanjiangyuan, the Reserve occupied 42% of Sanjiangyuan and it is an area with the most concentrated ecological species, the most important ecological position and the most complete ecological system in the entire region. Besides, it is the key to protect and govern the ecological environment of the Reserve and restore ecological functions of the Sanjiangyuan region. In 2005, the State Council approved the implementation of General Plan for Ecological Protection and Construction of Qinghai Sanjiangyuan Nature Reserve (referred to as the Plan). Accordingly, Sanjiangyuan ecological protection and construction project known as "No. 1 Ecological Project of China in the New Century" was officially launched with a total planned investment of RMB 7.5 billion. The Plan covered 22 projects under three categories, including 12 ecological protection and construction projects whose mission involves turning grazing land to grassland, turning farmland to forests (grassland), closing hillsides to facilitate forestation, desertification prevention and control, protecting wetland ecological system, comprehensive management of black soil land, preventing fire on forests and plains, taking precautions against rodent pests, maintaining water and soil, building management facilities and abilities, protecting wild animals and forbidding fishing in lake wetlands; 6 production and living infrastructure construction projects are built for farmers and herdsmen, including eco-migration, small town construction, supporting livestock raising project, energy construction project, irrigation and forage base construction and drinking water for people and livestock project; 4 ecological protection support projects, include artificial rainfall enhancement project, scientific research and application promotion, ecological monitoring and training of farmers and herdsmen.

Table 3.1  Contents of General Plan for Ecological Protection and Construction of Sanjiangyuan Natural Reserve

| No. | Category | Construction Contents |
|---|---|---|
| 1 | Ecological protection and construction projects (12 projects) | Returning grazing land to grassland, returning farmland to forests (grassland), closing hillsides to facilitate forestation, desertification prevention and controlpreventing deserted land, protecting wetland ecological system, comprehensive management ofgoverning black soil land, preventing fire on forests and plains, taking precautions against rodent pests, maintaining water and soil, building management facilities and abilities, protecting wild animals and forbidding fishing in lake wetlands |
| 2 | Production and living infrastructure construction projects for farmers and herdsmen (6 projects) | Ecomigration, small town construction, supporting livestock raising project, energy construction project, irrigation and forage base construction and drinking water for people and livestock project |
| 3 | Ecological protection support projects (4 projects) | Artificial rainfall enhancement project, scientific research and application promotion, ecological monitoring and training of farmers and herdsmen |

Up to the end of 2013, construction tasks in the Plan were fully completed and the tendency of ecological system degradation was preliminarily contained; the ecological state of key ecological construction areas were improved; the long-term nature and arduousness of ecological construction became prominent.

In 2008, opinions of the State Council about Supporting Economic and Social Development of Tibetan Regions in Qinghai and other provinces (General Office of the State Council [2008] No. 34 document) proposed "to timely launch early studies on Phase II project of Sanjiangyuan" and "establish the Sanjiangyuan National Comprehensive Ecological Protection Pilot Zone". In November 2011, the 181st executive meeting of the State Council approved the implementation of General Plan for the Qinghai Sanjiangyuan National Comprehensive Ecological Protection Pilot Zone, and required to compile special plans for key fields such as ecological protection, social cause, infrastructure, urban construction, industry and water resource. The comprehensive pilot zone is expanded to the entire Sanjiangyuan region based on the Reserve with a total area of 395,000 square kilometers. It covers 21 counties of the Yushu-Tibetan Autonomous Prefecture, the Goluo-Tibetan Autonomous Prefecture, the Hainan-Tibetan Autonomous Prefecture, the Huangnan-Tibetan Autonomous Prefecture, and the Tanggula Mountain Town of the Geermu City with 158 towns and 1,214 administrative villages. In order to carry out the central government's instructions of the Sanjiangyuan Ecological Protection and Construction, with the support and help of related central departments, Qinghai Government set out to plan the Sanjiangyuan Ecological Protection and Construction Project.

Qinghai organized the compilation of the Sanjiangyuan National Park System Pilot Plan in accordance with the National Park System Establishment Pilot Plan jointly issued by 13 departments including the National Development and Reform Commission, the State Commission Office of Public Sectors Reform, the Ministry of Finance, the Ministry of Land and Resources, the Ministry of Environmental Protection, the Ministry of Housing and Urban-Rural Development, the Ministry of Water Resources, the Ministry of Agriculture, the Ministry of Forestry, the National Tourism Administration, the State Administration of Cultural Heritage, the State Oceanic Administration and Legislative Affairs Office of the the State Council. On December 9, 2015, the general secretary Xi Jin-ping presided over the 19th meeting of the central deepening reform leading team and approved the Sanjiangyuan National Park System Pilot Plan. In March 2016, the General Office of the CPC Central Committee and the General Office of the State Council announced the Pilot Plan and the national park system pilot began. The Qinghai Party Committee and the People's Government regarded Sanjiangyuan national park system pilot as a No. 1 reform project and spared no effort to promote it. Besides, they set up a leading team with double leaders of the secretary of the the Provincial Party Committee and the Provincial Governor to formulate Deployment Opinions about Implementing "the Sanjiangyuan National Park System Pilot Plan", to organize a management institution and convene a mobilization meeting.

It officially launched the Sanjiangyuan National Park System Pilot Program and drew the prelude of China's first genuine national park construction. In order to ensure scientific advance of the national park construction, the leading team arranged compilation of the national park plan.

The Sanjiangyuan National Park includes the Yangtze River Source Park, the Yellow River Source Park and the Lancang River Source Park with an area of 123,100 square kilometers.

Conducting Sanjiangyuan national park system pilot is an important reform initiative of Qinghai Province to innovate its ecological protection management system and mechanism. Realizing "two unified executions" and changing the pattern of "multiple governance" are the core task and significant goal of the national park system pilot. The pilot will focus on the construction of the ecological protection management system characterized by clear jurisdiction, explicit responsibilities and effective supervision, and completely change the situation of Sanjiangyuan governed by different departments and industries with unsmooth system and implicit responsibilities. The Sanjiangyuan National Park will be built into a model area for ecological civilization system reform, thus exploring a new path that can be imitated and promoted for national ecological civilization construction. The Sanjiangyuan National Park System Pilot is another key initiative of Qinghai for ecological civilization construction which is elevated to a national strategy. Therefore, it is a milestone and it is significant to accelerate ecological civilization system construction of Qinghai and even the entire country.

## Section 2    Obvious Implementation Effect of the General Plan for the Nature Reserve

Since the implementation of the general plan for the nature reserve, departments of Qinghai and Sanjiangyuan at all levels have strengthened their leadership, optimized project layout, improved various systems, heightened project management, paid attention to science & technology support and carried out policies that can benefit people. The progress has been smooth. Within nine years from 2005 to 2013, 8.539 billion RMB in total is invested (excluding the subsidiary for plain ecological protection implemented after 2011), including a national investment of 6.588 billion RMB, local investment of 1.418 billion RMB and public self-investment of 533 million RMB, accounting for 114% of the planned investment. In order to effectively monitor and assess the ecological effect of the Sanjiangyuan Nature Reserve ecological protection and construction project, the Qinghai Department of Environmental Protection, the Department of Water Resources, the Department of Agriculture and Pasture, the Department of Forestry and the Meteorological Bureau etc. jointly form a Sanjiangyuan ecological monitoring working team. Under the organization and coordination of the Qinghai Department of Environmental Protection, the Institute of Geographic Sciences and Natural Resources Research, CAS, serve as the technical initiator to comprehensively apply technical methods such as ground observation, remote sensing monitoring and model simulation and jointly accomplish the task of "comprehensive assessment of ecological effect of the Qinghai Sanjiangyuan Nature Reserve Ecological Protection and Construction Project (Phase I)" regarding expected ecological project goals and regional ecological environment features on the basis of building comprehensive assessment indicator system and ecological background. The Institute of Geographic Sciences and Natural Resources Research, CAS and the Qinghai Sanjiangyuan ecological monitoring team conducted scientific monitoring and assessment of the ecological effect for nine years in succession based on the General Plan. The assessment showed that through nine years' arduous efforts of Qinghai and related central departments, "the ecological system degradation tendency of Sanjiangyuan has been preliminarily contained and the ecological situation of key ecological construction areas have improved." The effect is mainly reflected in five "increases": increase the vegetation coverage, increase the quantity of water resources, increase biodiversity, increase the income of farmers, herdsmen and increase social harmony. Goals established in the Plan are basically realized.

## I  The Ecological Effect was Significant, Mainly Demonstrated as Follows:

(I) The vegetation coverage is increased. From 2004 to 2012, vegetation coverage in the Sanjiangyuan Nature Reserve is rising. The forest coverage rate increases from 6.09% in 2004 to 6.99% in 2012; the plain vegetation degree has increased by 11.6 percentage on average. From 2004 to 2008, compared with the situation before, the implementation of the project, the area of the part with the same degradation status accounts for 69.35% of the original total degradation area; the area of the part with slight improvement accounts for 21.87%; the area of the part with obvious improvement occupies 7.40%; the area of the part with degradation accounted for 0.81%; the area of the part with intensified degradation is 0.57%. Fig. 3.1 indicates that the grassland degradation tendency is preliminarily contained and some degraded grassland tends to improve.

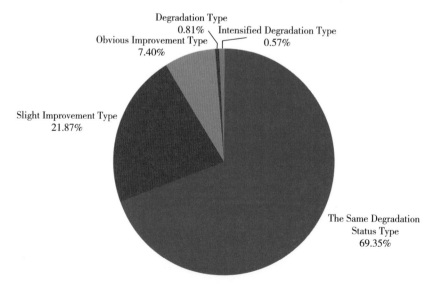

**Fig. 3.1  Proportions of degradation for the grassland of the Sanjiangyuan Region after the implementation of the Project**

(II) The runoff volume of rivers increases steadily. From 1975 to 2004, the annual average runoff of the Zhimenda Station in the Yangtze River is 12.43 billion cubic meters. From 2004 to 2012, the annual average runoff is 16.42 billion cubic meters with an annual average increase of 3.99 billion cubic meters. In 2015, the runoff is 15.502 billion cubic meters. From 1975 to 2004, the annual average runoff of the Tangnaihai Station in the Yellow River is 20.19 billion cubic meters; from 2004 to 2012, the annual average runoff is 20.76 billion cubic meters with an annual average increase of 570 million cubic meters; in 2015, the runoff is 15.801 billion cubic meters. From 2004 to 2012, the annual average runoff of the Lancang River is 10.7 billion cubic meters and it is 10.251 billion cubic meters in 2015. In 2015, the general condition of surface water quality in Sanjiangyuan is assessed as excellent.

(III) The water conservation function was apparently enhanced. From 1997 to 2004, the average annual water conservation amount of Sanjiangyuan forest-grass ecological system is 14.249 billion cubic meters and it is 16.471 billion cubic meters from 2004 to 2012, showing an increase of 2.222 billion cubic meters. The number of waters greater and equal to one square kilometers is 226 with a total area of 5,785.5 square kilometers, an increase

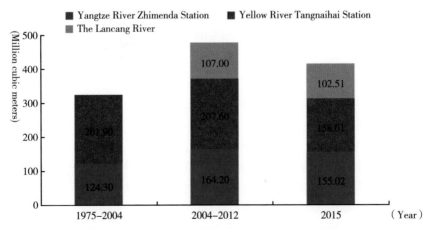

**Fig. 3.2 Statistical chart of changes in the net flow of the Yangtze River, the Yellow River and the Lancang River**

of 261.25 square kilometers over 2006. The areas of the Zhaling Lake and the Eling Lake have increased by 32.69 square kilometers and 64.36 square kilometers, respectively with a respective growth of 6.47% and 11.03%. In 2015, the average value of the water content of grassland soil in Sanjiangyuan is between 7.5% and 15.5%.

(IV) The ecological system structure gradually becomes benign. It is mainly demonstrated in expansion of the wetland ecological system, protection and restoration of the desert ecological system and contained degradation of grassland. From 2004 to 2012, the forest area of Sanjiangyuan has increased by 15.3 square kilometers and the grassland area has risen by 123.7 square kilometers while the water and wetland area has ascended by 279.85 square kilometers and the desert area has reduced by 492.61 square kilometers.

(V) Changes takes place in regional climate. From 1975 to 2004, the annual average temperature of weather stations in Sanjiangyuan is -0.58 ℃; the annual average temperature variation ratio is about 0.38 ℃/10 years. From 2004 to 2012, the annual average temperature of weather stations is 0.4 ℃ and the annual average temperature variation ratio is about 0.1 ℃/10 years, showing an obviously lowered temperature increase rate. From 1975 to 2004, the mean annual rainfall of stations is 470.62mm, and it is 518.66mm from 2004 to 2012 with an annual rainfall increase of 48.04mm and the moisture index has increased by about 5.03 on average.

(VI) The water and soil maintenance function is improved. From 1997 to 2004, the annual average soil maintenance amount of Sanjiangyuan is 546 million tons and it is 723 million tons from 2004 to 2012 with an increase of 177 million tons.

(VII) Biodiversity is increased. The quantity of wild animals such as Tibetan kiang, gharal and wild yak have increased obviously in the Nature Reserve and their inhabitant scope takes on an escalating trend. The diversity of plant population on the plateau and aquatic organism such as indigenous fish is effectively protected and biodiversity is gradually restored.

(VIII) The grazing pressure on the natural grassland is alleviated. Up to

2013, 3.42 million sheep units have been reduced for the natural grassland in the Nature Reserve. The livestock overloaded rate has lowered by 41.94 percentage. The grazing pressure is apparently mitigated. Besides, through eco-migration, 10,700 households and 53,900 people are transferred from the Nature Reserve, thus weakening the utilization of natural grassland by herdsmen.

(IX) The ecological prevention system and monitoring network are basically sound. The project zone's ecological prevention and monitoring abilities are apparently enhanced by establishing forestry, grassland and Natural Reserve guarantee system and strengthening ecological and environment monitoring network.

## Section 3　Phase II Plan of the Sanjiangyuan Comprehensive Ecological Protection Pilot Zone

The Qinghai Provincial Party Committee and the Provincial Government firmly carried out the decision made by the Central Party and the State Council, and regarded ecological protection as their important responsibility and the compilation of Phase II Project Plan of the Qinghai Sanjiangyuan Ecological Protection and Construction as a priority. The Phase II Project Plan, after being reviewed at the 100th executive meeting of the Qinghai People's Government and the fourth Standing Committee Meeting of the 12th Provincial Party Committee, was submitted to the National Development and Reform Commission (NDRC) in October 2012. The NDRC solicited opinions of the Ministry of Science and Technology, the Ministry of Finance, the Ministry of Land and Resources, the Ministry of Environmental Protection, the Ministry of Water Resources, the Ministry of Agriculture, the State Forestry Bureau and the China Meteorological Administration etc. and entrusted the China International Engineering Consulting Corporation to assess it. Revision and improvement were conducted according to the assessment report and opinions of different departments.

The NDRC approved Phase II Project Plan of the Qinghai Sanjiangyuan Ecological Protection and Construction on January 8, 2014 with a total planned investment of 16.057 billion RMB, covering 24 projects under two categories. At the same time, the Phase II Plan was officially initiated.

The Phase II Project Plan is the extension, expansion and enhancement of the Phase I Project, the main carrier of the General Plan and the major support for coordinating ecological protection, livelihood improvement and regional development. Although ecological degradation situation in the Sanjiangyuan National Nature Reserve has been improved, the vegetation coverage has been increased and biodiversity has been ameliorated since the implementation of the Phase I Project, there is still a long way to restore the ecological system function of Sanjiangyuan. The community structure, species composition and soil nutrients are still at the competition stage of evolution and degradation. Ecological and engineering measures should be further taken to consolidate ecological construction achievements. Hence, the implementation of the Phase II Project Plan is helpful to adopt more powerful measures and deali with ecological problems of Sanjiangyuan. The main contents of the Phase II Project Plan are listed below:

## I  Phase II Scope

The scope is the Qinghai Sanjiangyuan National Ecological Protection Comprehensive Pilot Zone, including 21 counties of the Yushu-Tibetan Autonomous Prefecture, the Guoluo-Tibetan Autonomous Prefecture, the Hainan-Tibetan Autonomous Prefecture, the Huangnan-Tibetan Autonomous Prefecture, and the Tanggula Mountain Town of the Geermu City with 158 towns and 1,214 administrative villages. It has a total area of 395,000 square kilometers.

## II  Phase II Goals

The base year of the plan is 2011 and the plan period is from 2013 to 2020.

Main goals: By 2020, the forest-grass vegetation will be effectively protected and the forest coverage rate will increase from 4.8% to 5.5%; the grassland vegetation coverage will increase by 25 to 30 percentage on average; land desertification will be effectively contained and governable desertification land governance rate will reach 50%; the vegetation coverage rate in the deserted land zone will be between 30% and 50%; the water and soil maintenance ability, water conservation ability and runoff stability will be strengthened; the wetland ecological system and wild animal and plant habitat environment will be obviously improved; biodiversity will be prominently restored; the production and living standards of farmers and herdsmen will be steadily increased; the ecological compensation mechanism will be further enhanced and the ecological system will enter a virtuous cycle.

## III  Functional Division of the Phase II Project

The General Plan divides the Sanjiangyuan Pilot Zone into key protection zone, ordinary protection zone and transfer undertaking zone in accordance with natural conditions, resource and environment bearing capacity, economic, social development and regional function orientation of the pilot zone. The divisional layout of the plan is consistent with the pilot zone. Based on the features of different divisions, space layout is optimized and categorized, instructions are conducted to ensure protected key technical points and main tasks so as to reach the general goals.

**(I) Key Protection Zone**

The key protection zone refers to the area which plays a specially important role in the ecological safety pattern of Sanjiangyuan. Ecological environment protection is regarded as the core, the development and operation activities are forbidden in principle.

*1. Regional Scope*

It focuses on the National Nature Reserve, including three national natural reserves of the Sanjiangyuan, the Kekexili and the Longbao National Nature Reserve. National geological parks, forest parks, wetland parks and scenic spots such as the Nianbaoyuze, the Kanbula and the Guide with an area of 1.98 million square kilometers, accounting for 50.1% of the total planned area.

*2. Regional Functions*

In this region, alpine steppe and meadow, glacier, marsh wetland, rivers and lakes are scattered with coniferous forests. Abundant and diverse wild animals and plants live in this region. The main function of this region is water conservation and production, and it is the core area of ecological protection. The ecological environment in this region is fragile. Once it is damaged, it will be extremely difficult to restore. Hence, according to related state laws and regulations, strict ecological environment protection measures must be taken to protect and restore.

**(II) Ordinary Protection Zone**

The Ordinary protection zone refers to the area which plays a fundamental role in maintaining integrity of the ecological system in the ecological safety pattern of Sanjiangyuan.

Priority should be given to ecological environment protection and animal husbandry should be properly conducted within the bearing capacity of resource and environment in this region.

*1. Regional Scope*

It covers areas beyond the key protection zone and transfers undertaking zone. It has a total area of 189,100 square kilometers, accounting for 47.9% of the total planned area.

*2. Regional Functions*

The region is the major area for alpine meadow, steppe and arid & semi-arid steppe with relatively rich biodiversity. It is a traditional region for production and operation activities such as grassland animal husbandry. It has certain resource and environment bearing capacities. Greater efforts should be made to protect and construct this region, control utilization of the steppe should be strengthened and balance between forage and animal should be realized; reasonable arrangements should be made to the pasturing area and guidances shoud be made to herdsmen; ecological system repair and restoration should be promoted, and ecological balance and harmony between man and nature should be achieved.

**(III) Transfer Undertaking Zone**

The transfer undertaking zone is the area which has certain potential of undertaking transferred farming and pasturing population and conducting industrial development, and it requires coordinated development and ecological protection in the process of ecological protection & construction and urbanization.

*1. Regional Scope*

It includes the Yellow River valley areas and the Gonghe County, the Guide County, the Jainca County and the Tongren County. Prefectures, counties and key small towns which are basically dotted distributed are also included with a total area of 8,000 square kilometers, taking up 2% of the total planned area.

*2. Regional Functions*

The ecological types in this region are complicated and diverse with overlapped farming and grazing, relatively concentrated farming land, desirable development conditions in the irrigated area with relatively high resource and environment bearing capacities. It is the main area where towns are developed, and industries and population are gathered. Under the prerequisite of ecological protection and construction, an urban system with prefectures and counties as the center, towns as the focus should be built to undertake the population transferred from the key protection zone and ordinary protection zone, and thus improving the environment and enhancing social security and service functions.

Transform development pattern, scientifically plan comprehensive development along the Yellow River and strengthen infrastructure construction for farming and animal husbandry; bring industries to parks based on actual conditions, develop characteristic and advantageous industries in order and make great efforts to achieve low-carbon and green development.

There are something missing in the table, for example, the Rigional Function if the Ordinary Protection Zone is not completed.

**Table 3.2  Functional division of the Phase II Project**

| Functional Division | Meaning | Regional Scope | Regional Functions |
|---|---|---|---|
| Key Protection Zone | The area which plays a special important role in the ecological safety pattern of Sanjiangyuan. ecological environment protection is rega-rded as the core, the development and operation activities are forbidden in principle | It focuses on National Nature Reserve, including three national natural reserves of the Sanjiangyuan, the Kekexili and the Longbao National Nature Reserve. National geological parks, forest parks, wetland parks and scenic spots Such as the Nianbaoyuze, the Kanbula and the Guide with an area of 1.98 million square kilometers, accounting for 50.1% of the total planned area. | Alpine steppe and meadow, glacier, marsh wetland, rivers and lakes are scattered with coniferous forests and abundant and diverse wild animals and plants. It mainly performs the function of water conservation and production and it's the core area of ecological function. The ecological environment in this region is fragile. Once it's damaged, it will be extremely difficult to restore. Hence, according to related state laws ard regulations, strict ecological environment protection measures must be taken to protect and restore it |
| Ordinary Protection Zone | The area which plays a fundamental role in maintaining integrity of the ecological system in the ecological safety pattern of Sanjiangyuan. Priority should be given to ecological environment protection and animal husbandry should be properly conducted within the bearing capacity of the resource and environment | It covers areas beyond the key protection zone and transfer undertaking zone. It has a total area of 189,100 square kilometers, accounting for 47.9% of the total planned area | The region is the major distribution area for alpine meadow and steppe and arid & semi- arid steppe with relatively rich biodiversity. It's a traditional region for production and operation activities such as grassland animal husbandry. It has a certain Capacity resource and environment bearing capacities. Greater efforts should be made to protect and construct it, control utilization of the steppe and realize the balance between forage and animal; reasonably arrange the pasturing area and guide herdsmen to cluster in order; promote ecological system repair and restoration and, and achieve ecological balance and harmony between man and nature |

*Continued*

| Functional Division | Meaning | Regional Scope | Regional Functions |
|---|---|---|---|
| Transfer Undertaking Zone | The area which has certain potential for undertaking transferred farming and pasturing population and conducting industrial development, and requires coordinated development and ecological protection in the process of ecological protection & construction and urbanization | It includes the Yellow River valley areas in counties such as Gonghe, Guide, Jainca and Tongren, and prefectures, counties and key small towns which are basically dotted distributed with a total area of 8,000 square kilometers, taking up 2% of the total planned area | The ecological types in this region are complicated and diverse with overlapped farming and grazing, relatively concentrated farming land, desirable development conditions in the irrigated area and relatively high resource and environment bearing capacities. It's the main area where towns are developed, and industries and population are gathered. Under the prerequisite of ecological protection and construction, an urban system with prefectures and counties as the center and towns as the focus should be built to undertake the population transferred from the key protection zone and ordinary protection zone, improve the environment and enhance social security and service functions; transform development pattern, scientifically plan comprehensive development along the Yellow River and strengthen infrastructure construction for farming and animal husbandry; bring industries to parks based on actual conditions, develop characteristic and advantageous industries in order and make great efforts to achieve low-carbon green development |

## IV  Construction Contents of the Phase II Plan

Main contents of the ecological protection and construction project are shown in Table 3.3.

**Table 3.3  Main contents of the Ecological Protection and Construction Project**

| No. | Category | Construction Contents |
|---|---|---|
| 1 | Protection and construction of the steppe ecosystem | Enclosure and grazing prohibition, and management of balance between forage and animal, return of grazing land to grassland, governance of black soil land and prevention & control of pests on the steppe |
| 2 | Protection and construction of the forest ecosystem | Management and protection of current forests, closure of hillsides to facilitate afforestation, artificial afforestation, update and transformation of agricultural protection forest, cultivation of young and middle-aged forests, construction of tree seedling base and prevention & control of pests in forests |
| 3 | Protection and construction of the desert ecosystem | |
| 4 | Protection and construction of the wetland, glacier, river and lake ecosystem | Water and soil maintenance, protection of wetland and snow mountain glacier, artificial intervention with the weather and protection of drinking water sources |
| 5 | Protection and construction of biodiversity | Fishing ban in lake wetland, fish proliferation and release and monitoring of endangered wildlife |
| 6 | Supporting projects | Ecological animal husbandry, rural energy construction, ecological monitoring, basic geographic information system, scientific research and promotion, training and publicity education |

## Section 4  Action of Community in the Society

The Sanjiangyuan ecological protection involves the ecological safety of China, Asia and the Northern Hemisphere, the long-term interest of the whole nation. With the general trend of global warming and abrupt ecological degradation, the Sanjiangyuan ecological environment faces a severe situation and has attracted great attention of the whole community. The Sanjiangyuan ecological protection is not only a concern of the country and the government but also closely related with every Chinese. It requires concerted efforts of

# The Course and Achievements of the Sanjiangyuan Ecological Environment Protection

the entire society.

In recent years, with social and economic development, people's awareness of environmental protection has been constantly strengthened. Besides, the enthusiasm of the whole community, varied groups and international organizations about participation in the Sanjiangyuan ecological protection, has been constantly enhanced. For instance, on March 12, 2001, Ericsson (China) Co., Ltd. donated transportation, communication and material equipment and a cash cheque of 2 million RMB for the protection of the Kekexili-Tibetan antelopes. Given the important ecological position of Sanjiangyuan, similar donations will be on the rise. In order to provide an open and reliable platform for domestic and foreign people who care about and support the public cause of the Sanjiangyuan ecological protection, the Sanjiangyuan Ecological Protection Foundation was officially set up on October 22, 2012. The Foundation is in strict compliance with related state laws and regulations as well as the mandatory rules of the Foundation. It has set up scientific internal management structures and rules of procedures. Besides, it conducts in accordance with Articles of Association of the Foundation, ensuring its standard operation and making good use of the money. It has become a powerful promoter of the Sanjiangyuan National Comprehensive Ecological Protection Pilot Zone. Since the inception of the Foundation, donations for the Sanjiangyuan ecological protection have been organized for many times to receive contributions from domestic and overseas organizations and individuals who are keen on the environmental protection cause. It unconditionally provides improved and standard services for donors, and fulfills donators' intention and applies donated capital effectively. Donations are mainly used in the Sanjiangyuan's resource conserving and ecological protection projects and scientific studies, and science & technology development, conducting and facilitating publicity education, academic exchanges and international exchange and cooperation regarding the Sanjiangyuan ecological protection. The Foundation makes sure that social funds can be fully invested in the Sanjiangyuan ecological protection cause. For instance, important projects such as "construction of benefit publicity platform", "proliferation and release of indigenous fish in the Yangtze River and the Yellow River" and "model area for comprehensive grassland governance" funded by the Foundation have obtained satisfying achievements in conducting publicity and education regarding the Sanjiangyuan ecological protection, disclosing related information, maintaining biodiversity, guaranteeing survival of endangered animals and plants, and comprehensively governing grassland ecology etc.

In addition to the Sanjiangyuan Ecological Protection Foundation, some self-organized public institutions such as the Snowland Great Rivers Environmental Protection Association etc. have made important contributions to the ecological protection of Sanjiangyuan. The Snowland Great Rivers Environmental Protection Association, formerly known as the Yushu Snowland Great Rivers Environmental Protection Association, was re-registered and renamed in the Qinghai Department of Civil Affairs on April 9, 2008. Members of the Association are mainly Tibetan people who are dedicated to protecting the ecological environment of the Qinghai-Tibetan Plateau and publicizing its traditional excellent ecological culture, and to pay attentions to sustainable development of the Qinghai-Tibetan Plateau.

Sanjiangyuan is a treasure land for scientific research. Concerning its particular location, it has formed unique and characteristic nature landscape and folk customs so it is of great value for scientific research. Meanwhile, scientific difficulties in the Sanjiangyuan ecological protection demand prompt solutions. In recent years, numerous universities and scientific research units have been engaged in scientific studies on Sanjiangyuan, and the number of academic papers and books about Sanjiangyuan has kept increasing with obvious achievements. On October 28, 2010, the Sanjiangyuan Research Institute of Qinghai University and Tsinghua University was established. It is initiated by Qinghai University and Tsinghua University to organize scientific research teams, which join hands with domestic and foreign universities and research institutions, and conduct comprehensive, systematic and targeted studies in such fields as the Sanjiangyuan ecological protection, following

# The Course and Achievements of the Sanjiangyuan Ecological Environment Protection

Fig. 3.3 The Sanjiangyuan Research Institute of Qinghai University and Tsinghua University

industrial development, sustainable development of public service system construction, comprehensive utilization of new energy and field station construction. At the same time, academic forums and seminars are held and research reports are published to boost restoration, protection and construction of Sanjiangyuan, and thus providing consultation services for governments. It is aimed to become a research institution that is well-known in China and influential to the world to some extent.

The Institute of Geographic Sciences and Natural Resources Research, CAS, set up a Sanjiangyuan ecology research team. Through eight years' study, it has referred to the theoretical framework of the UN's Millennium Ecosystem Assessment (MA), regarded assessment and effective continuous monitoring of ecological environment and assessment of the ecological construction as core goals, and spatial information technology as major means, and set up a comprehensive Sanjiangyuan ecological environment database system. In addition, it has designed and built a comprehensive assessment index system, developed an integrated ecological monitoring technology system of ground observation, ecological models and remote sensing observation, extracted rules of changes in the Sanjiangyuan ecosystem pattern and service functions since 1970s. It also has formulated a scientific, improved dynamic process background for assessing the effect of the Sanjiangyuan ecological environment project. Besides, the book of *Integrated Monitoring and Assessment of Sanjiangyuan Ecosystem* has been published through the effort in November 2012 which summarized and concluded key research achievements. Scientific studies on Sanjiangyuan provids a scientific basis for its ecological protection and restoration.

From June 20 to June 28, 2016, the Qinghai Department of Water Resources and Qinghai University jointly organized 27 professional technicians from Tsinghua University, the Qinghai Hydrology and Water Resources Investigation Bureau, the Huanghe Shuiwen Reconnaissance Design Institute, the Beijing Orient Landscape Co., Ltd., the Deqing Beichen Information Technology Co., Ltd., the Xi'an Summit Technology Co., Ltd. and the Tekhydro, Inc. in Beijing etc. to form the scientific investigation team of the Sanjiangyuan National Park.(see Fig.3.4)

Fig. 3.4　Scientific investigation team of the Sanjiangyuan National Park

# The Course and Achievements of the Sanjiangyuan Ecological Environment Protection

The scientific investigation team is aimed to strengthen hydrological monitoring of Sanjiangyuan, to explore hydrological change patterns, to improve the hydrological and water resource service ability and to provide technical support for the construction of "the Sanjiangyuan National Park" and "application of Qinghai Kekexili for world heritage" in a better way. The team set out from Xining, went through the Sanjiangyuan areas such as the Tongren, the Maduo, the Yushu, the Nangqian, the Zaduo, the Zhiduo and Qumarlai Counties, and carried out field surveys on the southern source of the Yangtze River— the Dang Qu River, its northern source—the Chuma'er River, positive source the Tuotuo River and the main stream—the Tongtian River. As well as the source rivers of the Lancang River—the Zha'a Qu, the Zhana Qu and the Zi Qu as well as current hydrological patrol stations of the Xialaxiu, the Xiangda and the Yanshiping, and gained basic data on the river system, weather, hydrology, riverway and river regime, water resources, water ecological environment, landform, social and economic conditions and communication conditions of the Yangtze River and the Lancang River source regions. Primitive information for site selection, reconstruction and upgrading of hydrological stations through field survey in the source regions was obtained, thus further improving and optimizing of the hydrological monitoring network can be achieved; investigation and learning about hydrological & ecological conditions of the source regions was carried out; examination of proper airborne water transfer area based on water-vapor flux studies was made. The investigation will be conducive to further understanding the ecological environment status quo of the Yangtze River source and the Lancang River source, thus providing scientific instructions for upstream development and protection of the great rivers.

In order to have a deeper knowledge and understanding of the Sanjiangyuan ecological environment conditions, from May 31 to June 7, 2017, followed by the scientific field trips since 2012, the Yangtze River Scientific Research Institute, the Qinghai Department of Water Resources and Qinghai University joined hands to conduct multi-disciplinary scientific investigation teams (being 34 members in total), and accomplished the multi-disciplinary comprehensive scientific investigation tasks focusing on the Geladaindong Snow Mountain and fully covering the hydrology and water resources, water ecological environment, river system, aquatic organism and bentonic organism, frozen soil and glacier, and landform of the positive source of the Tuotuo River of the Yangtze River, as well as the southern source's Dang Qu, the northern source's Chuma'er River's and the Lancang River's source. (see Fig.3.5)

**Fig. 3.5　Group photo at the Nangjibalong Scientific Investigation Monument**

# G.4 The Outlook of the Sanjiangyuan Ecological and Environmental Protection

The General Secretary Xi Jing-ping pointed out that the perils that water and the sources encountered are the same perils threatening the living environment and permanent existence of Chinese nation. We must value and conduct our struggle of solving the problem of water security on a high plane of constructing a well-being society and achieving a sustainable development. At the Third Session of the 12th National People's Congress, Premier Li Keqiang pointed out in the Government Work Report that we must doubly value the green properties granted to humans by the nature: forestry, grassland, rivers, lakes, and wetlands. To implement a good protection of Sanjiangyuan, we must enhance construction of major ecological programs, expand key ecological and functional zones, manage demonstration area well, develop comprehensive harnessing of territorial resources like rivers, and increase horizontal compensation between the upper range and the lower range. The new deployment strategies and demands have set up a new direction and a rare opportunity to speed up the work related to the Sanjiangyuan ecological and environmental protection.

The biggest values, the greatest potentials, and the toughest responsibilities of Sanjiangyuan lie in its ecology. This is a new beginning at a new historical point. There is still a long way to go to do well in ecological and environmental protection of Sanjiangyuan.

## Section 1  Challenges and Problems of Protecting Sanjiangyuan

We have gained some outstanding achievements periodically in the process of the Sanjiangyuan ecological and environmental protection. However, problems remain simultaneously. There is still a long way for us to go to protect and construct Sanjiangyuan.

Firstly, an overall degradation of ecology is not fundamentally curbed. With an initial protection and governance, the ecological degradation has been alleviated and changed to a better tendency. But there is still 80% black soil grassland and 60% desertified land out of control. 38% natural grassland is not executed under the policy of returning grazing land to grassland. Rats reemerge on the harnessed grassland. Within the whole area of Sanjiangyuan, intermediate degradated grassland occupies 50.4% of available area, among which, the black soil degradated type occupies approximately 39%; the desertified type occupies 8% of the planning area with an apparently increasing tendency; the soil erosion occupies 30.7% of the planning area; area with rats emergence reaches 167,000 square kilometers, accounting for 42.3% of the planning area. The increasingly degradating grassland, the desertified land area, and the area of the grasslands with pests and rats have greatly endangered the ecology of Sanjiangyuan. The general tendency of grassland degradation has not been fundamentally curbed.

Secondly, overgrazing still exists. The main factor leading to an ecological degradation of Sanjiangyuan is overgrazing. Previously, we have conducted policies of returning grazing land to grassland, constructing and breeding, and eco-migrating. However, due to the weak guidance of production transformation of farmers and herdsmen and the weak infrastructure construction of keeping livestock, the traditional grazing style continues. The main means of livestock reduction is mandatory. The fodder subsidies, grassland reward and subsidy mechanism can merely meet the basic life needs of the farmers and herdsmen rather than help them overcome poverty and achieve prosperity. That is the reason why the problem of overgrazing still exists. In Sanjiangyuan, there is 263,700 square kilometers natural grassland available, among which, over intermediate degradated grassland is 133,000 square kilometers; undegaded and slightly degraded grassland is 130,600 square kilometers. The actual carrying capacity is 18.2884 million per sheep. If we implement an overall grazing prohibition on the intermediate degradated grassland, then there is only theoretically 9.2663 million per sheep on the undegraded, slightly degraded, artificial pasture, and improved grassland. It thus occurrs as a prominent contradiction between the grass and the herds. And the solution to the problem is austere.

Thirdly, constraints of natural conditions and technologies are still serious. The average altitude of Sanjiangyuan is about 4,000m, with severe cold and thin air, strong ultraviolet, complicated terrain, unbalanced rainfall in time and space, vast frozen land, and slow soil and plant growth development. Once destroyed, it is extremely hard to recover. The special geographical conditions and fragile ecological features indicate that the work of carrying out an ecological recovery and management will be more difficult than that of the areas at a low altitude. Presently, ecological protection and recovering technologies of the natural environment against particular alpine region system are underdeveloped, thus leaving many virgin areas. In this case it restricts ecological protection and effects project construction. To be specific, there is only a single-method in treating black soil beach, wetland protection, wild animal protection, desertification control, and others. Although a series of scientific programs have been approved during the first stage, and the nation has arranged many supportive programs, some applying results are not obvious. For example, there are incomplete researches and underdevelped skills under site conditions in harnessing the black soil beach, planting artificial grass, and controlling rats. All these obstacles impede program construction.

Fourthly, the contradiction between ecological protection and livelihood improvement is prominent. As the continue of ecological protection and construction, the comprehensive demonstration area is expanded. The development and utility of mineral resources, water and electricity resources, Chinese-Tibetan herb medicine resources are constrained. Grazing prohibition and restriction area is increasing, which will definitely affect local economy development and living standard. This corresponds to a fact that the local population also rises. It is estimated that the number will increase to 1.34 million by 2020. The weak infrastructure cannot fulfill public service capacity. Industrial structure adjustment tends to be rather difficult. The ecological economy cannot form a short-term scale. There are many constraints in production and job transfer for the herdsmen. Besides, there is also an increasing contradiction between the rising population and the fragile environment capacity. Inconvenient traffic, difficult transportation, and short-term construction period increase both the difficulty of ecological construction implementation and the cost of construction. Therefore it is our major task and challenge to figure out an ecological protection and management system and an effective and standard long-term ecological compensation system which are beneficial to constructing a harmonious ecological protection, livelihood improvement, and social progress positively.

## Section 2   Carrying out the Sanjiangyuan Ecological Protection and Execution Regulations of the Phase II Program

As a subsequence to the expansion and a promotion of the first term program, "the Qinghai Sanjiangyuan Ecological Protection and Execution Regulations of the Second Phase Program" is a main support to overall ecological protection, improvement of livelihood, and regional development. After drawing the conclusion of the First Phase Program, we must focus on the following points in implementing the Second Phase Program.

Firstly, persisting in planning guidance under a uniform, general demand of ecological and cultural construction can lead the whole work of ecological protection. The stability of the Sanjiangyuan ecological system is closely related to a sustainable development of economy and society of the middle and lower ranges of the Yangtze River, the Yellow River, and the Lancang River. Practice has proved that the general plan has played a key role in promoting the Sanjiangyuan ecological protection and scientific development.

The Qinghai Provincial Committee and Government, taking the general demand of ecological and cultural construction as their guidance, have established a strategy of promoting comprehensive development of Qinghai Province through environmental protection. An idea of overall situation of building a national ecological security barrier is developed, following the guidance and basic principles of the general planning to reinforce leadership, to consolidate and concert different duties, to organize elaborately and implement smoothly. A solid foundation of achieving ecological system at Sanjiangyuan has been well laid.

Secondly, strengthening organization, leadership and execution can implement the planning, Qinghai Province has established a leading group and an executive office at three levels: province, prefecture, and county. Eight sub-groups have been put into effect, respectively, in charge of agriculture and husbandry, forestry, ecological migration, technical consultation, ecological monitoring, and others based on the nature of the projects. Within the program area, there is a grading contract system of responsibility involving county, township, village, and household, with the major leader of each prefecture and county as the primarily responsible party. These measures are the guarantee for implementing the program successfully. Next, the party committee and the government at each level must take the Sanjiangyuan ecological protection as a cut-in point to achieve implementation of promoting comprehensive development of Qinghai Province and changes in mode of development through environmental protection. We must strengthen leadership and full implementation of the work responsibility. We will take the Sanjiangyuan ecological protection and construction as a part of the performance evaluation of different governments to form a nice atmosphere to fulfill the task. Leaders of all levels must take the work seriously, interact and promote with joint efforts to proceed.

Thirdly, explore actively and enhance system innovation. We must abide by the relevant policies strictly in program management, program construction, ecological community management; as importantly, we must offer support to industry after ecological migration, and fulfill grass and animal balance mechanism. Meanwhile, we must explore actively and enhance system innovation to supply powerful support to the implementation of the program. For example, in the process of constructing the First Phase Program, we have initially established "Document of Probing and Establishing the Sanjiangyuan Ecological Compensation Mechanism", in which there are 11 items that greatly attract participation of the local people from ecological site. We have built up a Foundation of the Sanjiangyuan Ecological Protection Follow-up Industry Development and the Sanjiangyuan Ecological Protection Foundation, to make full use of market resources, to raise fund from multiple means, and to use intensively all funds to conduct resource compensation and ecological compensation. By doing so, we have carried out a co-ordinated settlement of ecological and environmental protection and construction, regional economic and social development, guarantee and improvement of livelihood, and achieved equalization of public service for local people. We have also adopted a green performance evaluation.

Fourthly, handing good relationship between protection and development can realize a consolidation of scientific protection and green developments. In the course of the Sanjiangyuan ecological protection, we must persist natural recovery as the primary means, while artificial harnessing as an auxiliary. We must strengthen management in dealing with irrational human behaviors such as overgrazing, grassland reclamation, unauthorized, wasteful mining, and illegal poaching. We will improve ecological system gradually by rat control, grassland fence, artificial grassland construction, natural grassland improvement, hillsides closure, afforestation facilitation, and wetland protection. We must stick to the policy of "protecting by developing and developing by protecting" to solve the easier problems first and the hard ones following to boost stably. We will implement all items related to the policies benefiting the local people. We will reinforce infrastructure construction of drainage, electricity supply, road, education, and hygiene by combining them with ecological protection program. We will support ecological migration followed-up industry with great efforts so as to ensure that the migrating people will move out, live comfortably, and become rich expectedly. By doing so, we will boost inner productivity in Sanjiangyuan and achieve a coordinated development between the ecological protection and the development of economy and society.

## Section 3  Further Strengthening Sicentific and Technological Investment and Research

The major part to enhance the Sanjiangyuan ecological protection lies in support ability of science and technology.

During the process of the First Phase Planning and Implementation, we focused on applied research programs which take ecological protection as the key task and extend their effects. There are 96 programs in total, among which, the findings of "Research on the Sanjiangyuan wetland changes and recovery technologies" and "Recovery and harnessment of the Sanjiangyuan ecological system of degraded grassland" achieved international advanced level.

We set standards for classification of the black soil beach, got a clear map of area, types and distribution of its degraded grassland. Treatment plan was proposed, and information system was built up. The initial causes of the formation of different degraded types were figured out and the recovery mechanism, sustainable countermeasures and feasible modes were put forward. The successful cultivation of the Festuca sinensis Keng and Qinghai Poa crymophila Keng provides significant support to harness degraded grassland. We set a good example of improving the Sanjiangyuan ecological monitoring and evaluating system continuously by improving the construction of the monitor station and team. As required by the program, we trained more than 6,000 persons/time of managers, approximately 50,000 persons/time of farmers and herdsmen, and constructed demonstrations for over 1,700 households. These measures are not just powerful support for the implementation of the programs. More importantly, we have trained a large number of managing staff, production-transfer and get-rich talents, hence increasing an overall promotion effect on the research findings, and have improved protection level.

In order to enhance further research and overcome key technical barriers of the Sanjiangyuan ecological protection, we have set new goals for the next step.

Firstly, advance ecological monitoring. We plan to improve five systems including ecological monitoring system, information transmission system, early warning service system, technique guarantee system for the monitor station system, and evaluating service system. We will construct monitor platforms such as the Ecological Environment Remote Sensing Analysis Center and the GOP Ground Observation Point. We will build up a database and information query system to improve information resource sharing, consultation and analysis mechanism. We will carry out a comprehensive report with an annual evaluation to uplift monitoring, warning, and appraising capacity. By doing so, we will have an overall and regular assessment of ecological environment circumstances, ecological system structure, ecological function, ecological sensitivity, carrying capacity of resources and environment, and ecological recovery effects.

Secondly, advance fundamental science research. This includes succession mechanism of the Sanjiangyuan ecological system, relation between climate change and ecological protection, and quantitative analysis of key index of the Sanjiangyuan ecology. We will have a better scientific understanding and judgment of the Sanjiangyuan ecological change. And thus we will get a clearer picture of its mechanism and propose corresponding treatment measures.

Thirdly, expand technical research and popularization. Research on key technical problems related to recovery of grassland, wetland, woodland, desertified land of ecological livestock husbandry, and organic husbandry, will help us to develop some advanced technique models of ecological protection, comprehensive ecological improvement, and industry development. We will strengthen skill training of herdsmen and technicians in a better way to popularize the skills over the Sanjiangyuan area.

Relevant departments under the State Council and all levels of government at Sanjiangyuan must adopt all possible and powerful measures to increase talent and capital investment to support scientific and technological research and application. This is a powerful support and guarantee to the Sanjiangyuan ecological protection and construction.

## Section 4   Constructing a Sanjiangyuan Protection System Involving Participation of the Entire Society

Sanjiangyuan is a place of different ethnic groups, with a majority population of Tibetan people, accounting for about 90% of the total population. People here have a pious belief in Tibetan Buddhism. The major industry is livestock husbandry which completely relies on ecological environment of the grassland. Tibetan people, with a nomadic culture and convention, have played a very important role in ecological protection since ancient time by making use of natural resources. This is beneficial for us to encourage the public participation, to enhance environmental protection awareness, and to set motivation model of active public participation.

Public participation can, on one hand, generate a common agreement on ecological protection and construction between the government and the public. It can, on the other hand, educate the public how to protect the environment by increasing their awareness and make their participation a conscious behavior.

The ecological protection does not restrict on the region itself, but more importantly it takes the water and ecology security of the middle and lower reaches of the Yellow River, the Yangtze River and the Lancang River, even the benefits of the whole nation into consideration. Then, how to guide people of different areas and fields to take an active part in the Sanjiangyuan ecological protection has the following concerns.

Firstly, Ecological protection must get support from NGOs, because they are different from units and individuals, thus exerting greater influence.

For instance, the Sanjiangyuan Ecological Protection Foundation, established in 2012, is a pure public welfare and public-raising foundation. It has played a huge role in promoting and safeguarding the Sanjiangyuan ecological protection by developing fund-raising, financial aid supplying to the Sanjiangyuan resource and ecological protection programs, and carrying out international exchanges and cooperation.

Secondly, public participation must be adopted by liberalizing policy. There must be participation at the initial stage of policy-making, at the implementing process, and at the afterward inspection through a thorough participation. The policy-making process aims at solving problems emerging in the Sanjiangyuan ecological protection and construction. These problems are closely related to local people's vital interests. Therefore, in the course of making policies and regulations, we must listen to the public suggestions to help us ensure the programs' success. The implementing process aims at solving inefficient governmental supervision. The purpose is to get the public into a coordinative supervision to create a greater efficiency. The afterward inspecting process aims at having the public joining in implementation of accountability in order to achieve the goals of ecological protection and construction in a better way.

Thirdly, a further publicity and education must be insisted. We will make a better use of broadcasting, television, movies, newspapers, internet, blog, wechat, and other means of public media to enhance the Sanjiangyuan ecological protection and publicity. By ways of multichannel and multiangle publicity, we will attract more attention from different sectors of the community, deepen the understanding of the significance of the Sanjiangyuan ecological protection, strive to create a good atmosphere of all possible support, enthusiastic help, and popular participation.

# References

[1] Qin Dahe. *Sanjiangyuan Ecological Protection and Sustainable Development*. [M]. Science Press. 2014.

[2] Georgraphy and Resources Research Institute of Chinese Academy of Sciences. *Comprehensive Monitoring and Assessing of Sanjiangyuan Ecological System*. Science Press. 2012.

# Acknowledgement

Qinghai Department of Water Resources
Qinghai Meteorologic Bureau
Qinghai Forestry Department
Qinghai Environmental Protection Department
Qinghai Hydrology and Water Resources Monitoring Bureau
Sanjiangyuan National Park Management Bureau
Qinghai Sanjiangyuan Ecological Protection and Construction Office

社会科学文献出版社　　**皮书系列**

## ❖ 皮书起源 ❖

"皮书"起源于十七、十八世纪的英国，主要指官方或社会组织正式发表的重要文件或报告，多以"白皮书"命名。在中国，"皮书"这一概念被社会广泛接受，并被成功运作、发展成为一种全新的出版形态，则源于中国社会科学院社会科学文献出版社。

## ❖ 皮书定义 ❖

皮书是对中国与世界发展状况和热点问题进行年度监测，以专业的角度、专家的视野和实证研究方法，针对某一领域或区域现状与发展态势展开分析和预测，具备原创性、实证性、专业性、连续性、前沿性、时效性等特点的公开出版物，由一系列权威研究报告组成。

## ❖ 皮书作者 ❖

皮书系列的作者以中国社会科学院、著名高校、地方社会科学院的研究人员为主，多为国内一流研究机构的权威专家学者，他们的看法和观点代表了学界对中国与世界的现实和未来最高水平的解读与分析。

## ❖ 皮书荣誉 ❖

皮书系列已成为社会科学文献出版社的著名图书品牌和中国社会科学院的知名学术品牌。2016年，皮书系列正式列入"十三五"国家重点出版规划项目；2013~2018年，重点皮书列入中国社会科学院承担的国家哲学社会科学创新工程项目；2018年，59种院外皮书使用"中国社会科学院创新工程学术出版项目"标识。

**权威报告·一手数据·特色资源**

# 皮书数据库
## ANNUAL REPORT(YEARBOOK) DATABASE

## 当代中国经济与社会发展高端智库平台

### 所获荣誉

- 2016年,入选"'十三五'国家重点电子出版物出版规划骨干工程"
- 2015年,荣获"搜索中国正能量 点赞2015""创新中国科技创新奖"
- 2013年,荣获"中国出版政府奖·网络出版物奖"提名奖
- 连续多年荣获中国数字出版博览会"数字出版·优秀品牌"奖

### 成为会员

通过网址www.pishu.com.cn访问皮书数据库网站或下载皮书数据库APP,进行手机号码验证或邮箱验证即可成为皮书数据库会员。

### 会员福利

- 使用手机号码首次注册的会员,账号自动充值100元体验金,可直接购买和查看数据库内容(仅限PC端)。
- 已注册用户购书后可免费获赠100元皮书数据库充值卡。刮开充值卡涂层获取充值密码,登录并进入"会员中心"—"在线充值"—"充值卡充值",充值成功后即可购买和查看数据库内容(仅限PC端)。
- 会员福利最终解释权归社会科学文献出版社所有。

卡号:996726993238
密码:

数据库服务热线:400-008-6695
数据库服务QQ:2475522410
数据库服务邮箱:database@ssap.cn
图书销售热线:010-59367070/7028
图书服务QQ:1265056568
图书服务邮箱:duzhe@ssap.cn

# S 基本子库
# SUB DATABASE

## 中国社会发展数据库（下设 12 个子库）

全面整合国内外中国社会发展研究成果，汇聚独家统计数据、深度分析报告，涉及社会、人口、政治、教育、法律等 12 个领域，为了解中国社会发展动态、跟踪社会核心热点、分析社会发展趋势提供一站式资源搜索和数据分析与挖掘服务。

## 中国经济发展数据库（下设 12 个子库）

基于"皮书系列"中涉及中国经济发展的研究资料构建，内容涵盖宏观经济、农业经济、工业经济、产业经济等 12 个重点经济领域，为实时掌控经济运行态势、把握经济发展规律、洞察经济形势、进行经济决策提供参考和依据。

## 中国行业发展数据库（下设 17 个子库）

以中国国民经济行业分类为依据，覆盖金融业、旅游、医疗卫生、交通运输、能源矿产等 100 多个行业，跟踪分析国民经济相关行业市场运行状况和政策导向，汇集行业发展前沿资讯，为投资、从业及各种经济决策提供理论基础和实践指导。

## 中国区域发展数据库（下设 6 个子库）

对中国特定区域内的经济、社会、文化等领域现状与发展情况进行深度分析和预测，研究层级至县及县以下行政区，涉及地区、区域经济体、城市、农村等不同维度。为地方经济社会宏观态势研究、发展经验研究、案例分析提供数据服务。

## 中国文化传媒数据库（下设 18 个子库）

汇聚文化传媒领域专家观点、热点资讯，梳理国内外中国文化发展相关学术研究成果、一手统计数据，涵盖文化产业、新闻传播、电影娱乐、文学艺术、群众文化等 18 个重点研究领域。为文化传媒研究提供相关数据、研究报告和综合分析服务。

## 世界经济与国际关系数据库（下设 6 个子库）

立足"皮书系列"世界经济、国际关系相关学术资源，整合世界经济、国际政治、世界文化与科技、全球性问题、国际组织与国际法、区域研究 6 大领域研究成果，为世界经济与国际关系研究提供全方位数据分析，为决策和形势研判提供参考。

# 法律声明

"皮书系列"(含蓝皮书、绿皮书、黄皮书)之品牌由社会科学文献出版社最早使用并持续至今,现已被中国图书市场所熟知。"皮书系列"的相关商标已在中华人民共和国国家工商行政管理总局商标局注册,如LOGO( )、皮书、Pishu、经济蓝皮书、社会蓝皮书等。"皮书系列"图书的注册商标专用权及封面设计、版式设计的著作权均为社会科学文献出版社所有。未经社会科学文献出版社书面授权许可,任何使用与"皮书系列"图书注册商标、封面设计、版式设计相同或者近似的文字、图形或其组合的行为均系侵权行为。

经作者授权,本书的专有出版权及信息网络传播权等为社会科学文献出版社享有。未经社会科学文献出版社书面授权许可,任何就本书内容的复制、发行或以数字形式进行网络传播的行为均系侵权行为。

社会科学文献出版社将通过法律途径追究上述侵权行为的法律责任,维护自身合法权益。

欢迎社会各界人士对侵犯社会科学文献出版社上述权利的侵权行为进行举报。电话:010-59367121,电子邮箱:fawubu@ssap.cn。

社会科学文献出版社